INTRODUCING QUANTITATIVE
GEOGRAPHY

A VOLUME IN THE ROUTLEDGE GEOGRAPHY
AND ENVIRONMENT SERIES
Edited by Michael Bradford

INTRODUCING QUANTITATIVE GEOGRAPHY

Measurement, methods and generalised linear models

Larry O'Brien

London and New York

First published 1992
by Routledge
2 Park Square, Milton Park,
Abingdon, Oxon, OX14 4RN

Transferred to Digital Printing 2004

Simultaneously published in the USA and Canada
by Routledge
a division of Routledge, Chapman and Hall, Inc.
29 West 35th Street, New York, NY 10001

Typeset by J&L Composition Ltd, Filey, North Yorkshire

British Library Cataloguing in Publication Data
A catalogue record for this title is available from the British Library.

Library of Congress Cataloging in Publication Data
O'Brien, Larry, 1957–
    Introducing quantitative geography : measurement, methods,
and generalised linear models / Larry O'Brien.
        p.   cm. — (Geography and environment series)
    Includes bibliographical references and index.
    ISBN 0–415–00465–9 (HB). — ISBN 0–415–07558–0 (pbk.)
    1. Geography—Mathematics.   I. Title.   II. Series.
G70.23.036   1992
910′.015′1—dc20   91–28683
                                                                      CIP

CONTENTS

CONTENTS

ILLUSTRATIONS

FIGURES

PRINTOUTS

TABLES

PREFACE

Quantitative geography is something of a minefield. No one is exactly sure how large it is, and no pathway through it is entirely safe. Yet many geographers find themselves having to venture into it, either because, as students, it is a compulsory part of their degree studies, or as graduate researchers, because it may provide techniques to help process data. The purpose of quantitative geography is the development of numeracy, and, in particular, with equipping geographers with the skills needed to collect, process and interpret geographical data sets. Without a sense of numeracy, many geographers are liable to be overwhelmed by the sheer volume of numbers which they may collect in the field.

Thirty years of technical developments have expanded the horizons of quantitative geography and have made it a difficult area to assimilate. Teachers and users alike find it difficult to keep track of the new techniques and styles of working which have been suggested. Furthermore, the proliferation of techniques has made it harder to encourage numeracy as the learning tasks have got more extensive and complex. These difficulties are particularly apparent in the statistical areas of quantitative geography where new fields of study have been developed and added to the armoury of geographical techniques. Categorical data analysis is an obvious example. Since the 1960s there have been major developments in analysing categorical data collected by survey, questionnaire and behavioural experiment. Accommodating these developments has been difficult because the new techniques fit uneasily into traditional course designs.

The purpose of this book is to show how some of these difficulties may be minimised. It presents a selection of statistical techniques which may be considered 'first cousins', and which can be shown to be natural extensions of the simple descriptive statistics taught to most geography undergraduates. Such techniques are members of the family of 'generalised linear models', and include many of the techniques usually associated with continuous data analysis, as well as many of the new categorical techniques. As with human families, it is often easier to remember shared characteristics than individual foibles, and so it is hoped that this approach will make learning easier.

This book began life several jobs ago somewhere between Bristol, Cardiff and Cheltenham. Its development has also taken in Durham, Oxford and Newcastle. Many people have helped me with it, often indirectly. My largest debt is to Neil Wrigley who, as supervisor of my doctorate, got me started in the first place. Nick Chrisman, Mike Blakemore and Ewan Anderson have also been major sources of help and enthusiasm. The following have provided more specific help: Janet Townsend and Barbara Harrell-Bond for useful tips on intensive research; Pion Ltd for permission to reproduce Figure 2 of Guy (1983b) as Figure 2.1; Peter Dodds for help with GIMMS; Kath Lund, Eileen Beattie and Lynne Martindale for typing; the Cartography staff in Durham and Newcastle; Mike Bradford for lots of advice on how to write a textbook; and Tristan Palmer, Alan Jarvis and Peter Sowden at Routledge for their patience. Mark McFetridge deserves special thanks for his expert handling of incompatible computers. The usual limits on responsibility apply.

This book could not have been written without the help of my wife, Ottilia, a microbiologist who has graciously put up with my dabblings in geography. Matthew arrived when the book was almost finished, which is just as well. It is dedicated to them both.

1

INTRODUCTION

1.1 OVERVIEW

Over the last thirty years there has been a major growth in the use of quantitative methods in geography. Most branches of the discipline have investigated their value as research aids, and some have been transformed by the experience. New subdisciplines devoted to the development of quantitative geography and spatial analysis have also been established. These have reinforced the value of quantification by creating new research opportunities for the discipline as a whole; opportunities which are increasingly being exploited in the form of applied research contracts between geography, commercial agencies and governments (Wrigley and Bennett 1981; Department of the Environment 1987).

This trend in geographical research has been reflected in teaching. Today, most geography courses available in UK or US universities and colleges include the teaching of quantitative methods in their design, frequently as compulsory or core units in their first or second year programmes. The content of such units varies, but, in general, statistical topics tend to predominate. Such a teaching programme might aim to cover the following topics from first principles:

1 the use of summary numerical measures to describe central tendency and dispersion in data;
2 the use of graphs, plots and other 'pictures' to look for patterns and relationships in data;
3 the use of inferential techniques to relate observed 'sample' data to an unobserved 'population';
4 the specification and testing of statistical hypotheses;
5 the use of an algebraic equation (a 'model') to account for the form of the patterns and relationships identified.

In the majority of cases, quantitative methods teachers are likely to assume that their students will have little or no previous statistical knowledge.

In designing quantitative methods courses for undergraduate geographers, the tendency has been to teach the 'basics' of statistical analysis by beginning with an outline of some of the terminology to be used and culminating with a consideration of the merits and potential uses of techniques such as correlation, the least squares linear regression model and the analysis of variance. These techniques are part of the intellectual inheritance of quantitative geography and reflect the strong links between geography and econometrics that have existed since the 1960s. However, statistical theory has not remained stationary over this thirty-year period. Many changes and refinements have taken place, and a considerable number of new analytical techniques have been developed. Some of these are directly relevant to statistical work in geography.

The developments in the methods available to analyse categorical data (alternatively termed 'discrete' or 'qualitative' data) illustrate these changes particularly well. Thirty years ago the techniques available for use with such data were extremely rudimentary and rather intractable, consisting of little more than a series of descriptive summary measures, *ad hoc* devices and 'rules of thumb' (Duncan 1974). The chi-square statistic was one of the more useful of these measures but it too was not without its difficulties. As Mosteller (1968: 1) noted towards the end of the 1960s: 'sometimes this approach is enlightening, sometimes wasteful, but sometimes it does not go quite far enough'. The essence of this comment, which many other commentators appear to have shared, is that these measures were often insufficiently powerful to analyse categorical data adequately. Today, the situation has changed and in addition to these measures, researchers are now able to employ a range of powerful statistical models which are similar in certain respects to least squares linear regression. These include the logit and probit models for proportions, and the hierarchical log-linear models for contingency table counts.

The development of these techniques and their incorporation in published geographical research poses a major problem for geographers. Many geographical data sets include categorical data, or data which have to be presented as categorical in order to meet the requirements of a legal, statutory or ethical obligation (e.g., the preservation of confidentiality). As a result, today's geographers need to be aware of techniques and models which may be useful for the analysis of such data. At the very least, a basic knowledge of the new developments is required in order to understand the increasing number of research papers which use them. However, in spite of this need, it seems that very few quantitative methods courses currently provide such training. Worse still, it seems that categorical developments are often only presented as specialities which are peculiar to specific branches of the subject (e.g., logit regression in urban geography).

Two factors would seem to account for this. First, it is the generally-agreed experience of quantitative methods teachers that statistics units are difficult to teach given the quantity and type of information to be presented. Anything which seems to make the job harder by extending the content is likely to be resisted. Second, many teachers believe that the level of expertise required to present these developments is high. The justification for this view seems to be that the categorical developments look and feel quite different from the traditional techniques. Not only are the notation schemes different, but so too are the rather more complex areas of estimation and inference. Because of these, many teachers see the learning curve for categorical data analysis as being particularly steep, and so have been deterred from making more than a superficial effort to tackle it. In the absence of an integrated approach to teaching quantitative techniques these objections would be insurmountable. But integrated approaches do exist, and these may allow programmes to be developed which can incorporate categorical developments whilst simultaneously covering linear regression and the other traditional topics.

This book presents such an integrated approach. It contains information on a programme of teaching which treats linear regression and the categorical models as special cases of a common family: the class of generalised linear models. This class has a number of attractive properties which make it valuable as an organising framework:

1 All members of the class may be written in a common, consistent notation which can replace the more distinctive, idiosyncratic notations normally associated with individual models.
2 The components of the model which reflect the associations within the data – the 'parameters' – may all be estimated using a common procedure: maximum likelihood.
3 The members of the generalised class are all developments of simple statistical relationships usually taught in introductory statistics courses.

By emphasising 'family features', quantitative methods teachers can obtain considerable economies of effort in their teaching. Furthermore, those approaching the subject for the first time should also experience considerable economies of effort in their learning. This is made all the more pertinent by the availability of the GLIM computer system which has been designed specifically to fit generalised linear models and is widely available.

1.2   WHAT ARE GENERALISED LINEAR MODELS?

The term 'generalised linear models' refers to a class of linear statistical models which share some important mathematical characteristics. Some of the class members which are in regular use in geography are illustrated in Figure 1.1.

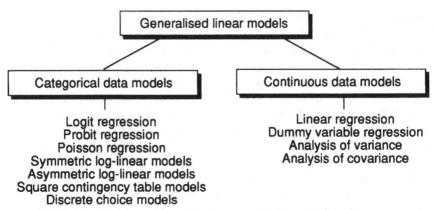

Figure 1.1 Some generalised linear models used in geography

From a teaching perspective, these models share three important characteristics. First, they may all be reduced to a common mathematical form and notation:

$$y_i = \mu_i + \varepsilon_i$$

where,

- y represents a 'response' component
- μ represents a systematic component which can be estimated from data and
- ε represents a random error component

Second, in each model the behaviour of the response is investigated by linking it to 'explanatory information' based on data collected by survey or experiment using some form of linear, additive structure. Third, the error component in each of these models is drawn from a common family of probability distributions known as the 'exponential family'. These three features distinguish the models in Figure 1.1 from the many other statistical models in common use, some of which are inherently non-linear in form, possess more than one error component, or have errors which are not from the exponential family.

The theory of generalised linear modelling was originally presented by statisticians J.A. Nelder and R.W.M. Wedderburn; their principal ideas being published in Nelder and Wedderburn (1972), Nelder (1974, 1977, 1984) and Wedderburn (1974a, 1974b, 1976). Other generalised frameworks have also been proposed by, among others, Scheffé (1959), Grizzle *et al.* (1969), Haberman (1970, 1974a), Horton (1978), Andersen (1980), Wrigley (1979, 1985). These contributions, though different in many important respects, lend further support to the idea that a generalised, integrated framework is particularly useful as a means of organising statistical material.

4

1.3 ORGANISATION OF THE BOOK

This book is organised into two parts, one introductory, the other rather more advanced. Part I (Chapters 2 to 6) is essentially an introduction to data analysis and covers the following topics:

1 A description of some of the sources and characteristics of geographical data.
2 The measurement of geographical information.
3 The description of this information using graphical and numerical summary measures.
4 The use of descriptive models to summarise geographical data which are distributed irregularly.
5 The use of statistical inference in geography.

The reason for including this is to provide a basis for the introduction of generalised linear modelling in Part II, by linking together the three important areas of data description, descriptive modelling and inferential modelling. These aspects are all features of numeracy and so should be emphasised in a data analysis text.

Part II (Chapters 7 to 11) is more advanced and illustrates the class of generalised linear models in more detail. Chapter 7 presents the background to the class and introduces notation, model specification and fitting conventions, and the maximum likelihood estimation strategy. Chapter 8 illustrates generalised linear models which may be applied to continuously distributed data – linear regression and the analyses of variance and co-variance. Chapter 9 illustrates generalised linear models suitable for analysing categorical data – log-linear, logit and probit models. Chapter 10 presents some additional types of generalised linear models which may be particularly valuable in geographical research, namely, Poisson regression models, log-linear re-expressions of spatial interaction models, linear experimental design models, capture–recapture models, and discrete choice models. Finally, Chapter 11 presents some conclusions and discusses possible limitations which restrict the usefulness of the class. Three appendices complete the book and cover some technical material on computer software, the probability background of the class, and the GLIM command language (versions 3.12 and 3.77), which is not presented in the text.

All of the examples presented in the book were tested using the mainframe version of GLIM (version 3.12) on Newcastle University's Amdahl computer, and on a microcomputer version of GLIM 3.77 running on an IBM-PC. Though GLIM has been emphasised, access to it is not a prerequisite as packages such as MINITAB, SPSSx, BMDP, SAS, GENSTAT, and GENCAT could be used instead, at least in part. These vary considerably in the demands they make of users, and in the character of their command languages. Some details of these packages are given in Appendix B.

Part I

PRELIMINARIES: GEOGRAPHICAL DATA AND DATA HANDLING

2

DESCRIBING GEOGRAPHICAL DATA

2.1 INTRODUCTION

A concern for numeracy is relevant to geography because the discipline is heavily dependent on data. A great deal of geographical research is based on the collection, analysis and interpretation of data, arising from sources as diverse as personal surveys and questionnaires, government publications, field measurements, laboratory experiments and satellite images. By 'data', geographers usually mean collections of facts and figures which refer to geographical 'individuals'. These 'individuals' may be distinct persons, and the data, observations on their ages, sexes, occupations or ethnic origins. Alternatively, the 'individuals' may be features in the landscape such as rivers, glaciers or peat bogs, and the data may be a series of observations on their dimensions, sinuosity, area or organic composition. However, just to add a little confusion, the term 'individual' may also be used to refer to social or physical aggregates such as zones, mountain ranges, drainage basins or travel-to-work areas. The data collected for these 'individuals' are actually aggregates of data gathered from sampling points within them. Thus unemployment in the geographical individual 'County Durham' is the total number of eligible unemployed claimants who are registered with Job Centres within the county.

The collection of geographical data can be extremely productive, with many thousands (even millions) of data items being involved in a single exercise. To process these quantities of data is practically impossible without suitable analytical tools. Computer processing is potentially helpful, especially in sorting and sifting the data. However, more valuable even than this is quantification, where the information contained in data is expressed in summary form by numerical techniques. Such techniques can be extremely rudimentary (means and variances – see Chapter 4) or rather more elaborate (regression models – see Chapter 8), with the choice of techniques reflecting the types of data being processed and the needs of the research.

The information content of geographical data varies with the type of data, how they were measured and collected, and how they were recorded

9

in the field. Similarly, the quality of the information reflects the assumptions of the collectors and the survey options open to them. These are important considerations because the definitions used in data collection 'depend on assumptions about the nature of meaning' (Sayer 1984a: 279), where 'meaning' reflects 'how we identify and differentiate the world' (ibid.: 280). In data sets collected by geographers for their own use, the quality of the data reflects the assumptions and preconditions which were thought most appropriate to meet the needs of the research. Because the users of the data set are also its collectors, they are in a good position to recognise its deficiencies and limitations. However, many geographers do not themselves collect the data they use for research, relying instead on the efforts of other researchers, government agencies or commerce to provide the data for their work. In these circumstances, the quality of the data is almost certainly less, as it is unlikely that the assumptions and world views of the collectors entirely match those of the users. Any analysis based on such data must recognise this basic fact, as no amount of clever numerical manipulation will add information to data where none already exists.

The purpose of this chapter is to review some of the characteristics of the data used in geography as a precursor to the more numerical material which follows. Five sets of contrasting terms are used to provide a method of organisation. These distinguish between:

1 primary and secondary data;
2 cross-sectional and time-series data;
3 'hard' and 'soft' data;
4 spatial and aspatial data;
5 categorical and continuous data.

These are not intended to be a rank ordering of geographical data types, nor are they meant to be exhaustive or mutually exclusive. Indeed, as the examples in this book will show, most geographical data sets consist of mixtures of data types, e.g., primary time-series, cross-sectional categorical data, 'soft' aspatial data and so on. However, they provide a recognisable summary of many of the major data types used in the discipline, and allow comments relevant to data quality to be advanced in a reasonably systematic way.

2.2 PRIMARY AND SECONDARY DATA

Collecting data is one of the more basic tasks of geographical enquiry. It is also one of the more arduous. Before any piece of research can commence, the researchers must decide what information is to be collected, and which sources and methods are to be used. As far as sources are concerned, a choice is often available. For example, they may be able to use existing data, even if collected for a different purpose, so long as they are sufficiently detailed, up-to-date, and can be adapted to the needs of the project. Alternatively, they may decide they have to collect new data.

The term 'primary data collection' is generally used to refer to the latter strategy, in which researchers collect information for a specific purpose 'in the field'. This may involve setting up an environmental monitoring system to collect information on daily temperature, rainfall or river flow. Alternatively, it may involve the use of a questionnaire, or other form of social survey, to obtain economic, social or attitudinal data directly from households or individual consumers (Marsh 1982; Bateson 1984). The main advantage of this approach is that the quality of the data is known and better understood, i.e., the researchers are better able to assess the effects of potential sources of error in the data because they have been intimately involved in their collection and recording.

Its main disadvantages are that it is often slow, arduous, labour intensive, and frequently expensive. It may take months, if not longer, to collect enough data of the type required for the study. Anderson and Cox (1986), for example, describe a study designed to assess the magnitude and velocity of soil creep at four sites in County Durham. Their data set, consisting of a series of daily readings, took nearly a year to collect. Similarly, Guy et al. (1983) and Wrigley et al. (1985) describe a study (known as the Cardiff Consumer Panel Survey) using self-completion shopping diaries which was designed to collect data on repetitive retailing behaviour and travel activity from a panel of several hundred households in Cardiff. This data set consists of many thousands of observations covering every shop visited, and grocery item purchased by the households, over a period of six months. Given its size, and the length of the survey, the data collection process was contracted out to a team of interviewers, specially recruited and trained by a professional market research company.

The delays and expense of primary data collection may be avoided entirely if a suitable existing data set can be adapted to meet the needs of the project. The obvious advantage of this approach (secondary data collection) is that the researchers can generally produce results more rapidly than if they have to collect the data for themselves. This may make their lives easier, but there is a risk involved, as the data are second-hand, and may therefore be of inferior or unknown quality. To be more specific, because the researchers may not have played any part in collecting the data, they may know very little about their quality, or about the effects of corrupting influences or errors in the data. An additional problem is that the information may be organised in unsatisfactory ways or refer to geographical areas which are not entirely appropriate for the research in hand. The suitability of the data to provide answers to the research questions being posed depends, in large part, on the seriousness of these inadequacies.

11

Some of the problems of secondary data analysis may be noted by inspecting the data in Table 2.1. This contains an extract of the migration flows into and out of Birmingham for both sexes and all ages for the period March 1987 to March 1988. The table shows a net loss of approximately 4,000 people from Birmingham over this period, with the largest outflow going to Solihull. The source of this migration data is the medical re-registration records of the National Health Service Central Register (NHSCR). This contains a record of all patient re-registrations with general practitioners in areas covered by Family Practitioner Committees. For the most part, these are shire counties or metropolitan districts. In London, however, they comprise combinations of boroughs.

Table 2.1 Migration (all ages) into and out of Birmingham to West Midlands (March 1987–March 1988)

| | Male | | | Female | | |
Area	In	Out	Gain	In	Out	Gain
Coventry	286	293	−7	339	312	27
Dudley	408	766	−358	461	847	−386
Sandwell	1,417	1,617	−200	1,632	1,812	−180
Solihull	1,887	3,023	−1,136	2,160	3,242	−1,082
Walsall	685	1,070	−385	897	1,235	−338
Wolverhampton	238	276	−38	257	313	−56
Totals	4,921	7,045	−2,124	5,746	7,761	−2,015

Source: NHSCR records (NOMIS)

The NHSCR is widely considered to be the best source of data for migration studies below the level of regions (Devis 1984). However, there are several problems which have to be recognised if these data are to be used for migration research. These are:

1 NHSCR data are not collected for migration purposes but for administering general practice medicine;
2 migration can occur without a re-registration taking place, especially if the patient is not a member of a 'high risk' medical category;
3 the speed of re-registration varies according to age and sex;
4 the areas used as spatial reporting units may not necessarily correspond to anything which is meaningful in socio-economic terms.

A number of studies (for example, Devis and Mills 1986; Boden *et al.* 1987) have shown that considerable discrepancies exist between the 'migration' patterns suggested by NHSCR data compared with Census of Population data. These are particularly noticeable for the migration patterns of adult males aged 20–29, potentially the most mobile group, who tend to be poorly represented by the NHSCR. This is because this group generally comprises healthy people whose medical requirements are

12

minimal and so who may not re-register as they move around the country. In the Birmingham example, only 345 out of the 4,000 or so movers were in this category. Other important mobility discrepancies are associated with the movement patterns of teenagers aged 15–19, pre-school children, and the elderly. For these groups, medical attention is probably quite important and so re-registration after each move is likely to be a high priority. In the Birmingham example, nearly half of the re-registrations were associated with children and elderly patients.

Apart from the National Health Service, other major sources of secondary data in the UK are government agencies such as the Office of Population Censuses and Surveys (OPCS), which collects and publishes the decennial Census of Population, and the Department of Employment which publishes unemployment statistics on a monthly basis as well as the triennial Census of Employment. Researchers may obtain this information in the form of printed records or, much more efficiently, using computerised information systems such as NOMIS (Townsend *et al.* 1987) or SASPAC (LAMSAC 1979, 1982; Rhind 1983). Local authorities are also useful sources of secondary data, providing considerable quantities of information specific to local areas. For example, the Electoral Roll, which they administer and maintain to determine who is eligible to vote in elections, may be used as a proxy list of the households living in specific streets or selected wards (see Guy *et al.* 1983 or Hoinville and Jowell 1978 for details). Similarly, libraries and other archives, such as County Records Offices, business directories and Trade Associations, may also be able to provide many types of secondary data which are valuable to the geographer. (For useful reviews of secondary statistical data, see Hakim 1982; Jacob 1984; Kiecolt and Nathan 1985; Durbin 1987; Openshaw and Goddard 1987; Department of the Environment 1987; Marsh 1989.)

2.3 CROSS-SECTIONAL AND TIME-SERIES DATA

One of the principal motivations for collecting data on geographical individuals is to highlight the properties and processes associated with them. These properties and processes may be related to each other in some systematic or organised way, operating independently or in conjunction to influence the characteristics of the individual. With time, they may also change their form, magnitude and relative influence. For example, the processes associated with landscape weathering may change over many thousands of years as a result of climatic changes. Similarly, trends in economic activity may change seasonally, or over the lifetime of a government, or over decades, as a result of new ways of working, technology and changes in the characteristics of international trade. At any one time, therefore, geographers may be interested in both the 'state' of a particular individual, and in how that state may change.

Table 2.2 County Durham: number of residents aged 16 or over present on Census night

Economic position	Total persons	Males		Females	
		SWD	*M*	*SWD*	*M*
All persons 16+	464,777	72,026	152,276	87,513	152,962
Econ. active	275,664	49,932	120,363	34,770	70,599
Working	239,815	38,461	104,894	29,935	66,525
Seeking work	31,733	10,611	13,296	4,339	3,487
Sick (T)	4,116	860	2,173	496	587
Econ. inactive	189,113	22,094	31,913	52,743	82,363
Sick (P)	15,396	3,393	7,091	3,231	1,681
Retired	48,871	10,275	23,751	10,110	4,735
Student	16,121	7,621	531	7,695	274
Other inactive	108,725	805	540	31,707	75,673

Key: SWD = single, widowed, divorced; M = married;

  Sick (T) = temporarily sick; Sick (P) = permanently sick.
Source: 1981 Population Census (via NOMIS) (see Townsend *et al.* 1987)

Geography frequently makes use of data which measure the state of a geographical individual at specific points in time. Such data are termed 'cross-sectional'. The Census of Population data displayed in Table 2.2 are an example of cross-sectional data. It shows the 'economic position' of all residents aged 16 or over in County Durham by sex and marital status on a single night in 1981. This information is specific to a single date – Census night (5 April 1981) – and may be found to differ from information on the same topics for the day before or after Census night. The main advantage of this type of 'frozen snapshot' is that considerable quantities of information are usually available for the chosen period. Thatcher (1984) notes that twenty-one questions were posed in the 1981 Census asking for data on household characteristics (e.g., on the size of each household, its age and sex composition, and on the dates of birth, marital status and countries of birth of each of its members), occupations and employment, journeys to work, qualifications attained, potential mobility, and on dwellings (e.g., on tenure, on the number of rooms, and on shared and exclusive amenities available to the household). Because this information was collected from almost every household in the country (approximately eighteen million in 1981) it represents the most detailed database available on the state of the British population in the early 1980s, containing excellent cross-sectional information for small areas (e.g., wards, parliamentary constituencies, parts of local authority areas).

  Time-series, in comparison, are generally much less detailed. Instead, they usually consist of a series of observations made on a geographical individual over a period of time (Chatfield 1980). The measurement of unemployment is an obvious example, consisting of a series of observations which are collected each month (Table 2.3). The main analytical use of time-series is in investigating the changes which occur through time in the state of the individual. The frequency with which observations are collected allows them to detect short-term, cyclical, periodic or seasonal changes in the state of the individual which cannot be detected by a cross-section. For example, information from monthly time-series of unemployment and vacancies in a seaside resort may be used to detect the seasonal patterns of employment change associated with the yearly holiday trade. In contrast, small area Census data will only be able to observe the employment and occupational structure of the town in 1971 and 1981, completely failing to detect patterns of change in the interim. Unfortunately, the cost and effort of collecting time-series data sets generally results in their lacking contextual cross-sectional information. This means that it is difficult to interpret why observed changes may have occurred.

There is, however, a hybrid form of data which possesses some of the advantages of both: 'panel' or 'longitudinal' data. Such data are usually collected by a special type of social survey – 'panel' or 'longitudinal' surveys – and are widely used in marketing, medical research, and in electoral opinion polling, to collect a detailed mix of cross-sectional contextual data

Table 2.3 Total number of unemployed claimants in County Durham

Date	Male	Female	All
Jan. 1985	30,719	12,215	42,934
Feb. 1985	30,062	11,973	42,035
Mar. 1985	29,608	11,818	41,426
Apr. 1985	29,948	12,017	41,965
May 1985	29,734	12,262	41,996
Jun. 1985	29,314	12,394	41,708
Jul. 1985	30,065	12,741	42,806
Aug. 1985	29,808	12,568	42,376
Sep. 1985	30,230	13,137	43,367
Oct. 1985	29,634	12,279	41,913
Nov. 1985	29,726	12,161	41,887
Dec. 1985	30,003	12,064	42,067
Jan. 1986	31,481	12,616	44,097
Feb. 1986	30,938	12,412	43,350
Mar. 1986	30,489	12,014	42,503
Apr. 1986	30,423	12,567	42,990
May 1986	29,678	12,384	42,062
Jun. 1986	28,810	12,200	41,010
Jul. 1986	28,812	12,362	41,174
Aug. 1986	28,584	12,169	40,753
Sep. 1986	28,889	12,683	41,572
Oct. 1986	27,690	11,865	39,555
Nov. 1986	27,658	11,436	39,094
Dec. 1986	27,522	11,314	38,836

Source: Unemployment statistics held on NOMIS

(e.g., on the socio-economic and former political allegiences of electors) and time-series behavioural data (see Ehrenberg 1972; Wrigley 1980; Wrigley *et al.* 1985; Davie *et al.* 1972; and Hamnett and Randolph 1987). The Cardiff Consumer Survey, which was introduced in section 2.2, is a recent geographical example of the type. An extract of its data is presented in Table 2.4 in the form of a number of computerised data 'records' coded up from the individual survey pages.

Table 2.4 Extract of Cardiff Panel data

Panellist	Time	Travel	Purchases
A01	31	14444115452	33,36
A01	32	24494114252	62,67
A01	33	34433114252	10,51
A01	34	44494114252	67
A01	34	54444115222	3,36,62
A01	35	60620215424	21,57
A01	35	60000000000	63,67
A01	35	77854115444	48,50
A01	35	84833115444	27,30,35
A01	35	94841115442	33,36,39
A04	32	14433111224	12,44,52

Source: Cardiff Consumer Panel Survey (Guy *et al.* 1983)

One of these was completed each time a household visited a shop to buy foodstuffs or other grocery items. Each record contains the following types of information:

1 General information on the household and where it is located in the city – digits 1–3 under 'panellist' (i.e., A01 is a household index).
2 Information on the timing of the shopping trip – digits 1 and 2 under 'time' (e.g., the first digit of 'time' refers to week 3 of the survey, the second digit refers to the day of that week, etc.).
3 Information on the shopping trip – digits 1–11 under 'travel', for example, which shops were visited, in what order and by what mode of transport.
4 Information on purchases made by the household, written as a variable number of digits under 'purchases'.

The value of the trip and timing information is its ability to place the shop visit in context. The record may be linked with preceding and subsequent records to describe daily shopping (e.g., household A01 made two separate shopping trips on day 4 but only one on day 1), shopping journeys (e.g., 'home–shop–home' and 'work–shop–work', see Guy 1983a), or aggregated over periods of one or more weeks to describe routine patterns of shopping behaviour. The household information allows the panel to be organised systematically into market types, allowing cross-sectional or time-series

16

analyses to be performed on representative aggregates of consumers. In turn, this allows the socio-economic composition of consumers in specified areas to be described in detail. Other possible analyses which could be based on this are the delimitation of market areas and spheres of influence, and a detailed examination of the dynamics of retail purchasing behaviour in the city or specified suburbs (see, for example, Wrigley and Dunn 1984a,b,c; 1985).

2.4 'HARD' AND 'SOFT' DATA

Given the variety of types of data and data collection processes available, it is natural to question which of these is most applicable for any given piece of research. The principal consideration here, should alternatives exist, is to ensure that data of an acceptable quality are collected.

The quality of geographical data is determined by a number of factors, some of which are essentially technical (reflecting the accuracy and precision of the measurements), while others are more subjective. A typical technical consideration is the ability to determine the likely margins of error associated with the data collection process. This is ably illustrated in socio-political research by opinion poll data which are collected to assess political preferences in a pre-election period. The usual practice is for a number of polls to be taken among 'representative' voters in different parts of the country, and the information they provide used to gauge the national feeling. However, because opinions are based on only a small number of voters in a limited number of areas, the political picture of the country as a whole is likely to be somewhat different from that recorded in any individual poll. This problem may be overcome to some extent by examining the margins of error around each poll (approximately 3 per cent in the polls conducted for both the latest General Election in Britain and the Presidential Election in the USA) to try to determine upper and lower levels of support for each party. This information provides a means of comparing the conclusions from each of the separate polls and estimating the levels of national support. Without this information, the individual results would probably be uninterpretable.

A similar problem also affects measurements made in physical research using monitoring stations or recording instruments. Such instrumentation must be appropriately calibrated for the environmental conditions in which it is operating, otherwise the results may be little more than rounding errors. Furthermore, though an instrument may be able to record numbers to many significant digits, the accuracy of these digits may be limited. Anderson (1988) notes that the Speedy Moisture Tester used in arid and semi-arid environments to record dewfall incidence is accurate to within 0.2 per cent. Figures from such a machine which purport to be more accurate than this are unlikely to be of very high quality.

Subjective influences such as religious convictions are also of importance in determining data quality. These manifest themselves at all stages in the data collection process, in the selection of study objectives, in the research design, and in the classification schemes and terminology used to describe the raw data. Terms such as 'class', 'poverty', and 'disability' illustrate the problem. Most social scientists regularly use terms such as these to refer to some state or characteristic of the world. However, it is unlikely that there is a single consistent meaning which may be applied universally to any of them. Many definitions may suggest themselves, and some are likely to be incompatible or mutually antagonistic.

To illustrate this point, consider the meaning of the term 'disability'. Blaxter (1976) suggests that two partially-incompatible definitions are paramount in British studies of the disabled. First, there is the approach which sees the disabled as 'innocent victims' requiring assistance, either from the state or charity. Such a view appears to have predominated in the social welfare studies in the nineteenth and early twentieth centuries when many of the associations for the blind and deaf were established. Second, there is the view that the disabled person is one who is:

> substantially handicapped in obtaining or keeping employment, or in undertaking work on his (sic) own account, of a kind which apart from his infirmity, disease or deformity, would be suited to his age, experience and qualifications.
>
> (Disabled Persons (Employment) Acts 1944–58)

The former characterises the disabled as persons whose bodies do not work correctly. The latter characterises them as persons who cannot work. Friedson (1965) goes further, by noting that disability is frequently seen as a form of social deviance. In this context, a person is disabled because he or she is seen to be a member of a stigmatised or deviant category; a category whose very existence reflects the commonly held, but rarely stated, perceptions of society about what is or is not 'normal'.

In recognition of this inherent problem of meaning (Sayer 1984a), one distinction which is sometimes used to describe the quality of geographical data is to note whether the information is 'hard' or 'soft'. In this context, 'hard' is used to refer to factual information which can be checked or verified in some way, whereas 'soft' refers to contextual or interpretative information such as opinions and attitudes. According to this distinction, population statistics are 'hard' because they are official 'facts' collected under statute. Unemployment statistics are also considered to be 'hard' for similar reasons. However, this does not mean that the 'facts' cannot change or be made to change. In the last few years the substance of these unemployment statistics has changed considerably. First, the data collection procedure used to measure unemployment has been altered from a monthly count of persons registering at job centres to a count of those

18

persons claiming unemployment benefit. Second, the types of claimant to be incorporated in the monthly count have also changed. Third, in early 1986, the date of the monthly count was switched from the second week in the month to the third. The cumulative effect of these changes has been to reduce the unemployment totals being published each month, and to generate considerable scepticism in some quarters about the quality of the published figures.

A second sense of the term 'hard' data is used to refer to data collection procedures which make use of the 'scientific' method developed for use in the pure and natural sciences. The primary purpose of this approach is to ensure that subjective bias (i.e., the bias introduced by the prejudices or idiosyncracies of the specific researchers) and other types of misinformation are minimised in research, allowing the results obtained to stand critical scrutiny by others. A major test of this is the ability to reproduce the results under similar experimental conditions. The same researchers, or more commonly, others located elsewhere, should be able to reproduce the stated results within a reasonable margin of error, merely by following the same procedures. If they can, the results assume the status of facts which are accepted by the research community. If they cannot, they will be regarded as doubtful by the profession, or ignored entirely as being of no interest.

In contrast to all this, the term 'soft' refers to data which are composed of non-verifiable or repeatable facts and figures. These may include opinions and beliefs, attitudes and prejudices, superstitions and fashions; indeed, anything which reflects people's perceptions of their own lives. It is often important to obtain contextual information on how people see their lives because behavioural actions are not pre-programmed but respond to perceived opportunities and threats. These may be impossible to quantify or verify with any reasonable accuracy, but may help to explain why two groups of people who appear to be identical according to hard data may behave very differently in response to a stimulus. The fact that hard figures cannot easily accommodate this type of behaviour indicates how poor they frequently are at representing the human environment.

Soft data abound in social research in, among others, questionnaire and preference surveys, ethnographic analyses, life-histories, geosophies and case studies (see Agar 1986, and Kirk and Miller 1986). They are frequently used in geography to study problems in which the definitions to be used are fuzzy – disability, part-time employment, 'second' jobs, urban – and objective, repeatable data are unlikely to be obtained. For further details on these and other fuzzy problems, see Williams (1973), Alden (1977), Harre (1979), Denham (1984), Pahl (1984), Sayer (1984a,b), Harrell-Bond (1986), Townsend (1986) and Gregory and Altman (1989). (A rather different use of the term 'soft' is considered in section 2.6.)

19

2.5 SPATIAL AND ASPATIAL DATA

The three types of data which have been described already are not exclusive to geography; many other subjects regularly make use of them to aid their research work. Geographers usually refer to these types as 'aspatial', because an explicit spatial or locational reference is not an integral part of the information they contain. In contrast, there is a data type which may be exclusively geographical: spatial data. Such data consist of observations on geographical individuals which may only be interpreted satisfactorily when their locations have been taken into consideration. This may involve a consideration of their absolute locations (site characteristics), or their relative locations as measured with respect to some benchmark such as the National Grid or sea level. Spatial data are often collected using maps, plans or charts (Unwin 1981; Mather 1991), but increasingly, they are to be found as the basic data type of computerised information systems, in which location provides an obvious and generally tractible method of organisation (Burrough 1986; Department of the Environment 1987).

By 'spatial', geographers usually mean data which are gathered in the form of points or dots, lines, areas or surfaces. However, the simplicity of this classification hides the fact that there are few generally accepted ways of describing, analysing and interpreting them. Haggett *et al.* (1977) suggest an approach involving the study of:

1 the distribution of points or 'nodes' in space and their relative positions with respect to each other;
2 the links or 'interactions' which appear to exist between them;
3 the man-made and natural 'networks' which funnel the interactions from node to node;
4 the 'hierarchies' which form within networks and between nodes which suggest the existence of some form of systematic human or physical organisation which may account for the form of the data;
5 the differing characteristics of the 'spheres of influence' which emanate from these hierarchies and which influence the spaces surrounding them.

This grammar is based on the locational analysis paradigm which is probably well known to most students of geography.

The information source for this type of analysis is often a dot map such as Figure 2.1, in which the retail provision in the eastern part of Reading is reduced to dots on a map. Each dot corresponds to an individual shop. By comparing the relative positions of the dots, it is possible for geographers to develop some appreciation of the spatial structure of retailing in the area, including, for example, the probable locations of shopping centres serving large market areas (the clusters) and isolated shops, or small groups of shops, serving local areas. It is also possible to detect the

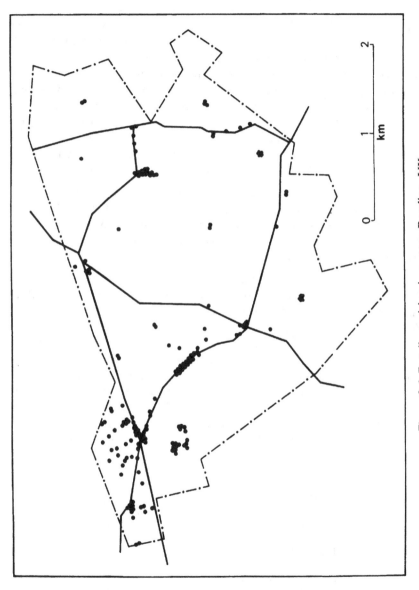

Figure 2.1 Retail provision in eastern Reading, UK

Reproduced from Environment cd Planning B; Planning cd Design, volume 10, 1983, p. 224, by permission of Piou, London

presence of ribbon shopping developments along transport routes. By incorporating this type of information with infrastructural and socio-economic information on, among others, the extent and state of the road network, car-parking provision, car-ownership rates, occupational profiles and family composition, researchers are able to assess the effectiveness of the provision and whether it meets the demands placed on it. Such 'accessibility' studies are of great interest to geographers and physical planners (see, for example, Massam 1975; Guy 1983b).

In such studies, accessibility is typically defined geometrically as representing the 'relative opportunity of interaction and contact' (Gregory 1981: 3). The effect of this definition has been to emphasise the importance of physical distance in spatial studies, and the costs of overcoming it. However, Harvey (1973: 57) notes that the costs of overcoming distance, and hence being accessible, are difficult to calculate, and include far more than geometric considerations. For example, he notes that the cost:

> can vary from the simple direct cost involved in transport to the emotional and psychological price imposed upon an individual who has an intense resistance to doing something (the kind of price which may be extorted, for example, from someone who has to take a means test to qualify for welfare).

This alternative perspective places the study of spatial structure within the wider study of social structure, and re-examines the meanings attached to terms such as 'distance' and 'proximity'.

A second form of spatial analysis is associated with zonal data. These are data sources in which disaggregate information from individual persons or landscape features has been aggregated and is displayed graphically by a system of geographical zones. Such data are frequently used to present the results of primary data collections, but they are also used in secondary data collections as spatial units of reference. Because the information in these reference units is an aggregation from the original sources, considerable care must be applied in interpreting it or using it in analysis.

Table 2.5 Zoning systems for use with GB population data

Zonal name	Number in GB
Individual wards	10,519
Local Authority Districts	459
Local Education Authority areas	116
Counties	66
MSC Training Division areas	58
Parliamentary constituencies	633
1984 travel-to-work areas	322
MSC/statistical regions	10/11

Source: NOMIS Users Manual, vol. 3 (O'Brien *et al.* 1987)

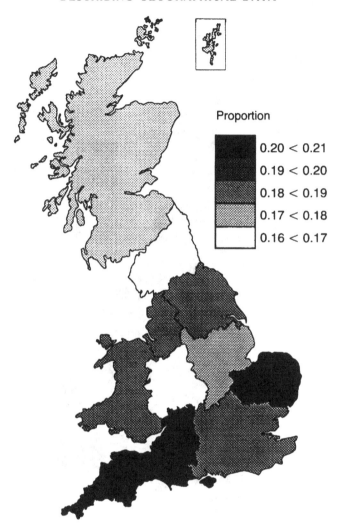

Proportion

0.20 < 0.21
0.19 < 0.20
0.18 < 0.19
0.17 < 0.18
0.16 < 0.17

Figure 2.2 Distribution of the elderly by region

To illustrate the basic idea, consider an analysis of the spatial pattern of the elderly in Britain. In 1981, the Population Census recorded that nearly one person in every six of the British population was elderly (approximately nine million out of fifty-four million). This information was collected from individual households located throughout the country and so may be displayed graphically as a map containing nine million dots. Though this idea is feasible, it is clearly impractical because the finished product would probably be unreadable. An alternative solution is to map the numbers of elderly persons by geographical zone. (Table 2.5 lists some of the zones available.)

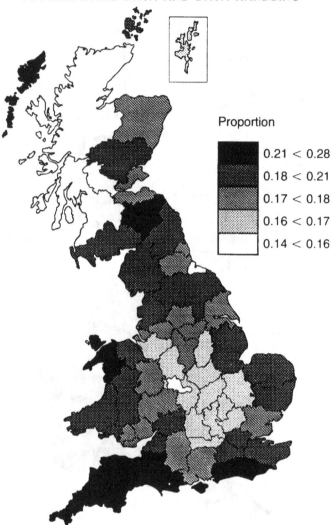

Proportion

■	0.21 < 0.28
▨	0.18 < 0.21
▨	0.17 < 0.18
░	0.16 < 0.17
□	0.14 < 0.16

Figure 2.3 Distribution of the elderly by county

Figures 2.2 to 2.4 present some maps of the elderly based on different zoning systems. The data being mapped are the porportions of elderly people in each zone. Each map gives rise to some very different spatial patterns. In Figure 2.2, for example, the zones with the highest proportions – the largest concentrations of elderly persons – are the statistical regions of the South-West and East Anglia. In general, the lowest areas are in the north and Midlands, the highest areas are all in the south. However, in Figure 2.3, which maps the elderly to county zones, the pattern is different. Instead of the whole of East Anglia, only the coastal counties

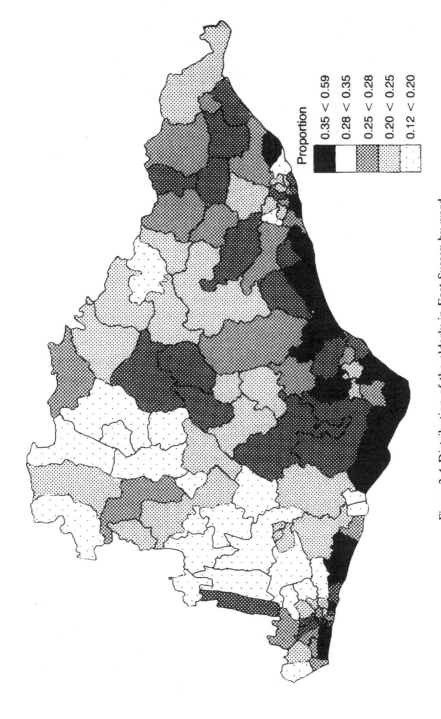

Figure 2.4 Distribution of the elderly in East Sussex by ward

Proportion

0.35 < 0.59
0.28 < 0.35
0.25 < 0.28
0.20 < 0.25
0.12 < 0.20

have particularly high proportions of elderly residents. Similarly, East Sussex and the other counties on the south coast exhibit very high proportions, even though, at region level, the South-East has a more moderate level. Moreover, as the ward map of East Sussex shows (Figure 2.4), the main pockets of the elderly appear to be located in coastal wards, rather than throughout the whole county.

These maps illustrate a key problem of zone-based spatial data analysis. Namely, that the maps are only able to 'measure the relationship between the variates for the specified units chosen for the work. They have no absolute validity independently of those units, but are relative to them' (Yule and Kendall 1950: 312). In other words, the dependency relationships being presented by the maps are not unambiguous facts; they are by-products of the zones which have been selected for the graphics. The reason for the apparent differences in the maps merely reflects the fact that the ratio of elderly to total population is not constant across the country but varies. In particular, its value reflects the choice of boundary for each zone. Though the pattern of this variability is suggested by the three maps, it should be noted that these are only three alternative perspectives. Given that there are nine million elderly persons located in the country, it follows that there may be as many maps. Which of these is the best?

The choice of a 'best' map is difficult to determine and depends on the needs of the research. Clearly, as one moves from the 11-zone region map of Figure 2.2 to the 66-zone county map of Figure 2.3, the complexity of the pattern increases as more boundaries are added. By supplying more boundaries, the number of ratios to be created increases. As these boundaries are irregular and the new zones differ in shape and size, this process is likely to increase the variability within the data and so generate an even more complex series of patterns.

In these figures the areas being used are Census zones defined for use with the 1981 Population Census. These are general-purpose spatial systems which are widely used to classify and present government statistics. However, there is no reason why any of these should be used to represent the spatial structure of the elderly or, for that matter, any other spatially-varying phenomenon. 'Off-the-peg' spatial systems can easily introduce distortions which make the interpretation of spatial data difficult and hazardous. Ideally, graphics should be based on classification procedures which have been devised specifically for the data being studied. This being the case, the question of suitability changes from one merely concerning the number of zones, to one concerning the number and type of zones.

This problem of measured relationships changing with the type and number of zones used is frequently termed the 'modifiable areal unit problem'. Its implications have been recognised for many years (see, for example, Yule and Kendall 1950; Openshaw and Taylor 1979; Openshaw

1983), and have led researchers to suggest a number of novel solutions which might minimise its effects. Two of these are the use of areal weighting or regular hexagons. The latter is an attempt to produce a recording unit in which the relationships being presented are invariant to the zone used. As such, it represents one attempt to identify a 'basic spatial unit' which can stand as a fundamental level of geographical investigation (Department of the Environment 1987). It is not surprising to find that there is no general agreement on what such a unit should be.

One of the main problems associated with the lack of a basic spatial unit is that zonal data can be distinctly misleading. Because the information recorded in the zone reflects both the disaggregated data gathered from persons within it and the denominator effects associated with the imposition of boundaries, it is possible to make wholly unjustified claims about the area and its inhabitants. For example, the ward data in Figure 2.4 suggest that the wards with the most elderly are coastal. This interpretation arises because the proportions calculated for those areas are relatively high. However, it is quite wrong to go on to state that all such areas are therefore inhabited exclusively, or even predominantly, by the elderly.

Table 2.6 Total number of unemployed persons in the northern region by travel-to-work area (1984 definition)

Zone name	Total
Newcastle upon Tyne	48,891
Sunderland	28,818
Middlesbrough	21,582
South Tyneside	11,246
Stockton-on-Tees	10,880
Hartlepool	7,185
Morpeth and Ashington	7,062
Durham	6,821
Bishop Auckland	5,746
Darlington	5,429
Carlisle	4,092
Barrow-in-Furness	3,439
Workington	3,346
Whitehaven	2,893
Alnwick and Amble	1,816
Hexham	1,037
Kendal	971
Berwick-upon-Tweed	746
Penrith	731
Windermere	244
Keswick	199
Column totals	173,174

Source: NOMIS (Department of Employment data)

Such an assertion would be an example of an 'ecological fallacy', as even in these wards, the non-elderly population make up between one-third and two-thirds of the population. The figures are thus representing a characteristic which is almost certainly a minority feature in most places.

In addition to the interpretational problems posed by modifiable areal units and the ecological fallacy, researchers need to be aware of how the data are represented in the map. To illustrate this consider the data in Table 2.6, which are the total number of unemployed persons in the travel-to-work areas in the northern statistical region in August 1988. The figures are listed in descending order, with Newcastle upon Tyne having the highest unemployment level and Keswick the least. Assuming that the zones to be used have already been chosen, the next choice to be made is the method of assigning the raw data to the shading levels used in the map. Table 2.7 lists twelve different 'default' procedures available in two popular computer mapping packages, GIMMS (Carruthers and Waugh 1988) and MAPICS (MAPICS 1986).

Table 2.7 Twelve methods of automatic classification used in GIMMS and MAPICS

METHOD	MAPICS	GIMMS
Equal arithmetic	1	EQUAL
Equal rounded arithmetic	2	ROUNDED
Curvilinear progression	3	CURVE
Geometric progression from zero	4	GEOZERO
Geometric progression of class widths	5	GEOWIDTH
Arithmetic progression of class widths	6	ARITH
Equal intervals on reciprocal scale	7	RECIP
Equal intervals on trigonometric scale	8	TRIG
Normal percentile	9	PERCENT
Proportional to standard deviates	10	STD
Quantile-based intervals	11	QUANTILE
Nested means intervals	12	NESTED

Note: To select a method of classification, use the CLASS command in MAPICS, or the TYPE= option in the GIMMS *INTERVALS command

These methods vary in the way they treat the raw data, and consequently, in the way they portray its information. The procedure of equal arithmetic, for example, calculates the largest and smallest figures in the data and subdivides the difference between them by the number of shading levels to be used. In this case, the largest figure is 48,891, the lowest is 199, and the number of levels is 5. Each level thus corresponds to 20 per cent of the data values. The 'equal levels' map of these data is displayed in Figure 2.5. Apart from five zones in the east, all areas are allocated to the bottom level in the map, indicating that their unemployment levels are within the bottom 20 per cent of values in the data set. The five areas

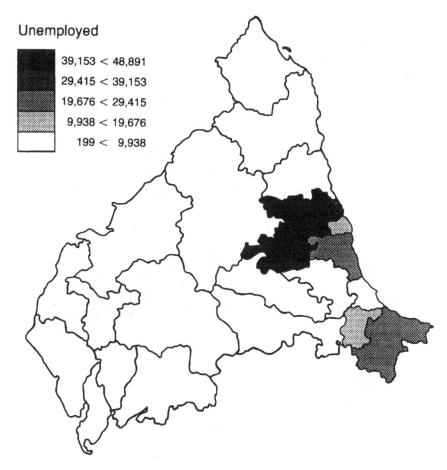

Figure 2.5 'Equal levels' map of data in Table 2.6

allocated to the other levels include the main population centres of Newcastle, Sunderland and Middlesbrough, and are traditional unemployment blackspots.

In contrast, the quantile procedure (method 11), calculates the data values for each level in such a way that approximately 20 per cent of the zones are allocated to each class (Figure 2.6). Thus the lowest level may be interpreted as corresponding to the 20 per cent of the zones in which unemployment is lowest, the top quantile to the 20 per cent in which unemployment is highest. This map breaks up the homogeneity of Figure 2.5. In particular, it distinguishes between the central Lake District zones, which have low unemployment, and the industrial zones along the Cumbrian coast which have higher unemployment. The map also discriminates between the more populated, industrial travel-to-work areas in

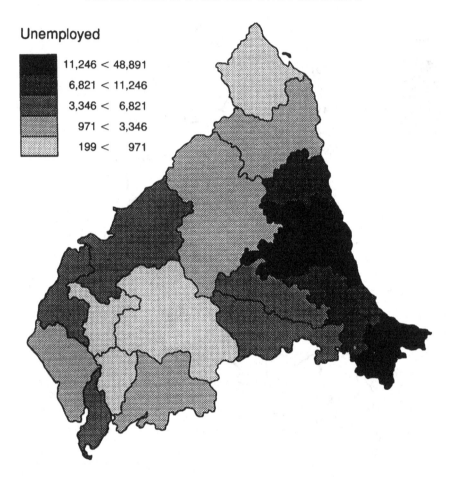

Unemployed

■	11,246 < 48,891
▨	6,821 < 11,246
▧	3,346 < 6,821
▤	971 < 3,346
▫	199 < 971

Figure 2.6 'Quantile' map of data in Table 2.6

Cumbria and the east, and the mainly agricultural and tourist areas in the Lake District, Hexham, and Northumberland (Berwick).

As a third example of this representational problem, consider Figure 2.7 which classifies the raw data using an arithmetic progression of class widths (method 6). The classification procedure is rather more complex than those used in the other maps and involves dividing the range of the data values by 15 (the arithmetic progression based on five levels, in which the second interval is twice as wide as the first, the third is three times as wide, etc.). This map divides the northern region into two distinct parts. To the west and north, the mainly agricultural and tourist areas are picked out as having relatively low levels of unemployment. The only exception here is the Carlisle area. To the east, all the more traditionally industrial areas are picked out. Most of these, however, are allocated to the second lowest level, indicating the dominance of the traditional population centres in determining the major patterns in the map.

2.6 CATEGORICAL AND CONTINUOUS DATA

The final distinction listed in section 2.1 contrasts data which are said to be categorical with those which are continuous. This is an important distinction in statistical analysis as many of the techniques which will be presented in the remainder of this book are based on it. The key difference between the two types is that continuous data are divisible, and categorical data are not.

Sex is an important categorical variable which is used to subdivide a population on the basis of chromosomal differences. The distinction into

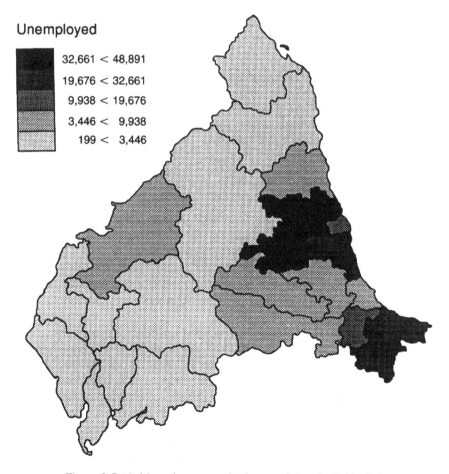

Figure 2.7 'Arithmetic progression' map of data in Table 2.6

31

male and female is straightforward, and provides a way of investigating whether features in the data are related to sex-group. A variable recording this would thus have two mutually exclusive levels: 1 (or M) for males, and 2 (or F) for females. Rainfall, on the other hand, can be measured to parts of centimetres, for example 1 centimetre, 0.2 centimetres, 1.45 centimetres. This ability to record information as real numbers rather than as whole, indivisible units characterises continuous measurements.

As will be shown in the next chapter, it is normally possible to measure all types of geographical variable as categorical, but only some as continuous. To illustrate this idea, consider the measurement of a physical variable such as dewfall in an arid environment. A simple way of gaining an impression of the temporal variability of dewfall incidence is to define an incidence variable (INC) and use it to record those days on which dewfall occurs and fails to occur. To distinguish these two states, days without dewfall could be coded 0, and those with could be coded 1. The sequence:

0 1 1 1 0 0 1 1 0 1 0

thus shows that dewfall occurred on six days, and failed to occur on five other days. In this example, the numbers 0 and 1 are merely codes distinguishing between the two states of INC. They have no mathematical significance apart from this. Letters (Y, N), words (YES, NO) or symbols (*, %) could have easily been used instead to record exactly the same information, namely:

N Y Y Y N N Y Y N Y N
* % % % * * % % * % *

However, if the research requires it, it is possible to record dewfall more accurately. Wherever INC equals 0 (or N or *), dewfall will be 0 millimetres. However, on days where INC is 1 (or Y or %), dewfall will be any non-negative real number, for example 1 millimetre, 0.245 millimetres, 3.4 millimetres. Sex, however, cannot be treated in this way as it makes no sense to record an individual as being 0.75 male or 1.34 female. Sex can only be measured in categories; its information is interpretable only in terms of categories. However, the dewfall incidence variable, INC, is a 'user-defined' categorical variable. The term 'user-defined' is used here to reflect the fact that the classifications are not natural or necessarily obvious, but are created by researchers to meet specific research needs.

In describing geographical data it is valuable to be aware of these three distinct types of measurement – fundamental categories, user-defined categories and continuous measurements – as their information content varies. Some of the implications of these distinctions will be taken up in Chapter 3, and in Part II in connection with the analysis of data tables. However, it is also likely to be of importance in mapping when the areal units are actually fundamental rather than user-defined categorisations (unlike the examples in the previous section). Chrisman (1981) refers to such maps as 'categorical coverages'.

2.7 SUMMARY

The variety of data types used in geography reflects the interests and concerns of geographical research. The quality of geographical information is similarly variable and reflects what the data are, and how they are measured, recorded and collected. Each of these stages is prone to error, not just in the measurement of the data (for example, using cross-sectional data as surrogate time-series), but also in their definition. As definitions reflect the predilections of the researchers, no data can be wholly objective. Indeed, one of the most productive avenues of geographical research is concerned with investigating how knowledge of the world is created and shared. A particular concern for the value of categorical measurements has been generated out of this interest.

Quantitative procedures provide one way of describing the information in geographical data sets which are of particular value if the number of items to be processed is large. However, the methods which are used must be appropriate to the data to avoid misrepresentation. Whilst it is possible to record all geographical variables as categories, only some of these may be re-expressed as continuous variables. As many of the more important numerical techniques developed for use in geography assume continuous data, it is likely that they will only be of occasional use. Techniques for categorical data have been developed since the late 1960s which allow geographers to perform sophisticated numerical analyses on this important data type. As the structure of Part II is based on the categorical–continuous distinction, it is important to consider what these measurement types mean in more detail. This is the objective of Chapter 3.

3

MEASURING GEOGRAPHICAL DATA

3.1 INTRODUCTION

The final distinction made in Chapter 2 to describe some of the types of data used in geography may also be used to describe their measurement. By 'measurement' geographers usually mean two things: first, the description of what the data represent (i.e., a 'naming' function), and second, the calculation of their quantity (i.e., a 'counting' function). Measurement thus includes the assignment of names to features on the ground, and the calculation of complex relationships using them, e.g., the assessment of the density of land use in a given area. The motivation for this is to provide a mechanism which may be used to describe geographical individuals and to communicate this description to others. Both aspects are most readily achieved using a 'language' which is at once unambiguous, flexible, plausible and generally accepted by researchers.

Numbers, and systems of numbers, offer a possible language which would seem to satisfy these requirements. According to Skemp (1971), numbers possess the following desirable properties:

1 they are intimately linked to the concept of counting, which is a pre-mathematical skill possessed by most people and which is a commonly-used method of evaluating the properties of a collection of objects (see also Piaget 1952);
2 they provide a precise mechanism for comparing objects on the basis of these properties, which may be either tangible, abstract, or both;
3 they allow objects to be grouped into classes or sets which share similar (or identical) characteristics. Such categorisation is often the basis of measurement;
4 they may be used to create new objects from the originals.

In other words, the measurement of the environment involves the categorisation of the observed data into distinct groupings which may be named (e.g., areas of the earth's surface may be labelled 'deserts' on the basis of annual rainfall), manipulated to create new objects (e.g., population density from population size and area), and new categories (e.g., areas of high, medium and low population density).

Measurements based on the use of numbers are said to be 'derived' measurements (Ellis 1966). This is because the structure of the data is inferred from patterns and relationships observed between the numbers. The assumed relationship between these patterns and the environment is a matter of some controversy. The so-called 'representational school' (e.g., Campbell 1928; Stevens 1946, 1951, 1959) appear to believe that there is a one-to-one correspondence between them. This implies that the structure in the numbers exactly mirrors that in the environment. As a result of this, they urge caution in the choice of numbers used in description, as an inappropriate choice will lead to confusion and misinformation (i.e., the patterns in the numbers will not correspond to anything which exists in the environment). In comparison, the 'pragmatists' (e.g., Adams 1966; Baird 1970) argue that no such correspondence exists, and that numbers merely provide a shorthand method of encapsulating information. As a result, different sorts of numbers should be used to 'mine' the data, without a particular type being chosen beforehand.

In spite of the gulf between these two 'schools', and the clear difference in their conceptions of the use of numbers, aspects of both are generally accepted in quantitative geography. The main support for the representational school lies in the use of 'scales of measurement'. In practice, geographers readily recognise four such scales – the nominal, ordinal, interval and ratio – which are assumed to govern the quality of the information which can be recorded from the environment. At the same time though, geographers frequently 'break the rules' associated with these scales (e.g., treating ordinal as interval) in an effort to search for deeper meaning in the data.

Table 3.1 Scales of measurement

Scale	Defining relationships	Permissible techniques	Geographical examples
Nominal	Equivalence	Non-parametric	Areal names Land use categories
Ordinal	Equivalence Magnitude	Non-parametric	Urban hierarchies Social class
Interval	Equivalence Magnitude Ratio of two intervals	Non-parametric Parametric	Centigrade or Fahrenheit
Ratio	Equivalence Magnitude Ratio of two intervals Ratio of two observations	Non-parametric Parametric	Population size Population density Distance Precipitation

Source: Adapted from Siegal (1956) and Unwin (1981)

Each scale is defined by Stevens (1946, 1951, 1959) using a series of defining relationships (see Table 3.1). These are assumed to correspond to (unstated) laws of nature, and are organised hierarchically. This means that only the ratio scale (the most advanced) exhibits all the defining operations while those further down the hierarchy possess fewer. Thus a ratio scale may be re-expressed as an interval, ordinal or nominal scale. The reverse (re-expressing a nominal as a ratio) is not possible. These scales may be related back to the data types used in geography in a direct way: the nominal and ordinal scales are usually used to measure categorical data; the others to measure continuous data.

3.2 CATEGORICAL SCALES OF MEASUREMENT

3.2.1 The nominal

The nominal scale is the simplest of the four scales recognised in Stevens's hierarchy. It is used principally to classify raw data into mutually exclusive sets or levels on the basis of characteristics which the researchers think relevant. The names used to describe land use provide a ready example of this type of measurement. For example, the First Land Utilization Survey in England and Wales (Stamp 1948) used six sets to represent the land surface. These were areas of built-up land, horticulture, cropland, grassland, woodland, and heath and unimproved land. Parcels of land were allocated to one of these six sets on the basis of their physical appearance. If a field appeared to the surveyors (many of whom were schoolchildren) to be mainly used for arable agriculture, it would be allocated to set 3 (cropland), this set being the land use category thought most appropriate of the six to represent the field.

The number and choice of sets depends on the purpose of the research. Stamp's study was mainly concerned with enumerating rural activities at a fairly general level. More recent studies have been concerned with urban activities, and have used a larger and more sophisticated choice of sets, based both on physical appearance and on functional characteristics, e.g., patterns of commuting (Coleman 1961; CURDS 1983; see also Rhind and Hudson 1980). However, regardless of the purpose of the research, two objectives motivate the definition of sets in nominal measurement. First, they should provide the researchers with an ability to classify every observation in the data set, leaving none unclassified. Second, they should be mutually exclusive. This means that none of the observations may be allocated to more than one set, e.g., land which is 'cropland' cannot simultaneously be classified as 'horticulture'. The property which allows observations to be handled in this way is the logical property of 'equivalence'. In terms of Stamp's study, two parcels of land are said to be equivalent if their principal land use is the same, e.g., both woodland areas.

The notion that observations are equivalent does not, however, imply that they are identical. The fields in the 'woodland' set may all be different sizes and the number of trees, or the percentage tree cover, in each may also be different. Other examples of observations which are equivalent but not identical are Dover and Southampton in the set 'English ports', and Durham and Canterbury in the set 'English towns with Anglican Cathedrals'.

Because nominal measurements only allow equivalence comparisons to be made they have tended to be ignored, or overlooked, in quantitative research in geography. The main attitude to them seems to have been that their information is impossible to manipulate mathematically. For example, the sets may be identified by names or a numerical key may be used:

1 to refer to towns containing cathedrals but not castles;
2 to refer to towns containing castles but not cathedrals;
3 to refer to towns containing both;
4 to refer to towns containing neither.

These numbers are for identification only and do not imply that the sets may be used in mathematical operations. It makes no sense to subtract set 2 from 4, or to add set 1 to 3. Similarly, the placing of a town in a specific set does not mean that it possesses only one of the distinguishing characteristics. For example, Durham and London would both be classified in set 3, without further account being taken of the fact that Durham only contains one cathedral whereas London contains several. However, even though the numerical keys cannot be manipulated mathematically, it is possible to count the number of observations allocated to each set and identify which of them is the most frequently occurring in the data. This provides very basic numerical information, but since the mid-1960s, numerical techniques have been available which can manipulate it mathematically, often in ways which offer a deeper insight into its structure.

3.2.2 The ordinal

The second of Stevens's four scales, and the second categorical scale in the hierarchy, is the ordinal. This provides a slightly more sophisticated form of measurement than the nominal because, in addition to equivalence, it allows the sets to be placed into some form of rank order (the logical property of magnitude). To illustrate this, consider the classification of topography into 'highlands' and 'lowlands'. The definition of these two classes incorporates information on height above sea level so that it is possible to state that 'highlands' are at higher altitudes than 'lowlands'. In terms of the defining property – height above sea level – 'highlands' exhibit more of it than 'lowlands'. They are thus qualitatively ordered.

There are many examples of ordinal classifications in geographical research. For example, 'development' can be used to classify countries according to their perceived economic health into categories labelled 'developed', 'developing' and 'underdeveloped'. Similarly, precipitation may be used to classify climate into groupings such as the equatorial westerly zone, which is constantly wet, the extratropical westerly zone, which has precipitation throughout the year, and the high polar zone, which has meagre precipitation (Flohn 1950). Third, Haggett *et al.* (1977) note that the road network of a country can also be organised in a rank order, with unclassified roads at one end, and motorways at the other. In each of these examples, some form of ranking has been used to arrange the classifications into a qualitative order.

The analysis of ordinal data is particularly important in survey research in which personal behaviour or opinions are being assessed. By asking shoppers to respond to a wide range of questions on shopping habits and trends, consumers can be targeted for advertising campaigns and promotions which are likely to interest only a proportion of the population (Harris and O'Brien 1988). The questions listed in Table 3.2 typify those used in these surveys. They are a subset of the 28 questions asked in the attitude questionnaire part of the Cardiff Consumer Survey (Guy *et al.* 1983). The responses to these questions were classified using a five-level coding – agree strongly, agree, neither agree nor disagree, disagree, disagree strongly – where each code is assumed to lie along an underlying agreement continuum.

Table 3.2 Example of Cardiff Panel Survey attitude questionnaire response categories

| Questions | Response categories | | | | |
	1	2	3	4	5
Going grocery shopping gives you the chance to meet friends	15	175	47	183	30
Given a choice between good shops and good parking, I would choose to shop where there is better parking	21	131	95	171	22
I would prefer to do all my shopping just once a week	37	188	23	186	15
I like to buy all my groceries at one shop rather than shopping around	38	169	24	194	25
Getting shopping done quickly is very important to me	54	197	39	153	7

Key: 1. Strongly Agree 2. Agree 3. Neither agree nor disagree
      4. Disagree 5. Strongly disagree
Source: Extracted from Table 5.16 of Cardiff Consumer Panel Survey (Guy *et al.* 1983)

These five sets differ from those used in the First Land Utilization Survey in that they are qualitatively related. Each represents a different level of agreement. Consumers prompted by the attitudinal questions are thus able to record the strength of their agreement with the prompt by selecting a set which most closely matches their feelings. It is thus possible to say that set 1 represents a larger (or stronger) degree of agreement with the prompt than set 2, which, in turn, represents a larger degree of agreement than set 3, and so on.

As before, the equivalence of these responses does not imply that they are identical. Consumers who allocate their responses to the same set may do so for completely different reasons, and, indeed, may interpret the question from a wholly different perspective. In other words, the knowledge that panellists in general agree with prompt 1 tells us nothing about why any particular panellist is in agreement. Similarly, magnitude does not imply that the differences in the level of agreement between the sets are equal, or that the scale is symmetric around the middle set. Indeed, ordinal scales provide no information at all on the relative 'distances' between the sets. They may, for example, be equally spaced along some continuum of agreement (Figure 3.1a), or they may be markedly skew with most of the sets gathered to one end (Figure 3.1b). Some consequences of this for the design and manipulation of preference questionnaires are described in Payne (1951), McKennall (1977), Presser and Schuman (1980) and Kalton *et al.* (1980).

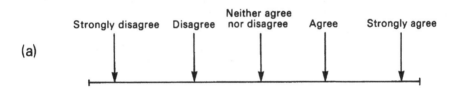

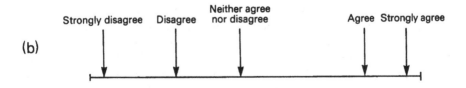

Figure 3.1 Examples of an ordinal scale

3.3 CONTINUOUS SCALES OF MEASUREMENT

Lying immediately above the two categorical (or qualitative) scales in Stevens's hierarchy are the two continuous (or quantitative) scales: the interval and ratio.

3.3.1 The interval

The interval is the cruder of the two continuous scales. It lies just above the ordinal in the hierarchy, and differs from it in that the distances between the sets are known. This extra piece of information is particularly valuable and allows researchers to be more precise in their reporting of the characteristics of the data.

Interval measures are characterised by their ability to class data items into sets (the equivalence property), place them in some form of rank order (the magnitude property), and describe the precise distances (the 'intervals') between them. These properties allow interval measures to overcome the problem illustrated in Figure 3.1 where the exact, quantitative relationships between sets is not known. Because these distances are known in interval measurement, it is possible to calculate how much one set differs from another. Thus if the interval between sets A and B equals that between sets B and C, it follows that the interval between A and C is twice that between A and B. Similarly, if A–B is half the interval between B–C, then A–C is three times A–B. This sort of processing allows the ratio of any two intervals to be calculated, a feature which is required if the sets are to be used mathematically. However, such ratios are restricted to the sets and do not apply to the observations within them (except in the special case where each set only contains a single observation). In other words, though A–C is three times A–B, it does not follow that an observation classified in set C has a value three times that of an observation in set A.

Interval scales are frequently used to record attitudinal and opinion information. A typical example of their use is in medical self-assessment surveys, where patients are asked to record their impressions of their medical condition before, during and after treatment. This is often done using a survey form such as Figure 3.2, in which patients are presented with a series of lines drawn horizontally across the form (representing an assessment continuum) which correspond to two or three time periods in the treatment of their condition. The patient records his/her feelings by marking a series of vertical lines somewhere along the horizontal lines. The distance separating these vertical lines from the origins of each horizontal line represents the strength of their assessments. These assessments may be compared by noting the relative position of these vertical lines (i.e., an improvement due to treatment should be detected by the gradual movement of the vertical lines towards the right-hand end of the

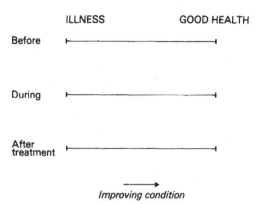

Figure 3.2 Example of an interval scale

scale). However, additional mathematical processing of the responses is limited because the origins are arbitrary rather than real, i.e., the zero point does not mean that the patient does not have an opinion on his/her medical condition.

A geographical modification of this approach is given in O'Brien (1982). This study attempted to measure the strength of residential preferences of a selection of council tenants who had recently moved to the council housing estates of Horfield, Lockleaze and Manor Farm in north-east Bristol. Movement within this housing market is heavily managed but, in spite of this, tenants (including prospective tenants) are given an opportunity to make known any preferences they may have for locality and house type. The sophistication of these preferences among the tenants in the study varied from:

1 an articulate locational preference in which the tenants identified a specific dwelling, to
2 a situation where no locational preferences were made, either because the tenants did not answer the preference question on the housing application, or because they stated that anywhere in the city would do.

Between these extremes the motivation for the requests varied, and included requests to live in specific areas irrespective of accommodation type (i.e., the primary motivation was for the area), as well as requests for specific types of dwelling irrespective of area (dwelling motivation). To display this information as an interval scale, each area which could be mentioned on the application needs to be given an index number to distinguish it from others. By combining these into sets (e.g., all applications with specific dwellings stated, all applications with specific housing estate mentioned, all applications with areas north of the river mentioned etc.), it is possible to measure the strength of the stated preferences and also check how frequently they were satisfied. The origin in this linear scale corresponds to the second of the two extreme types of preference; the upper end to the first. Though the origin is recorded as 0 in Table 3.3 this does not mean that the tenants did not have residential preferences in spite of none being mentioned. The origin is therefore arbitrary rather than real, a feature which characterises interval measurement.

Table 3.3 Codes for creating a geographical interval preference variable

Code	Interpretation
0	No preference made
1	Preference for any form of accommodation or location
2	Preference for north of the River Avon
3	Preference for housing management areas in same part of city
4	Preference for contiguous housing management areas
5	Preference for specific housing management area
6	Preference for contiguous estates
7	Preference for specific estate
8	Preference for specific part of estate
9	Preference for specific road
10	Preference for specific property

Source: Adapted from O'Brien (1982)

3.3.2 The ratio

The second of the two continuous scales, and the most sophisticated scale in the hierarchy, is the ratio. This scale differs from the interval in that ratios may be made between individual observations in the data set and not just between the sets. A second difference is that the origin of the scale (the zero point) is real and not arbitrary. This means that a measured value of zero really implies the absence of the phenomenon being studied. For example, zero distance means no distance travelled and zero precipitation means that none has been recorded. As a result of this, ratios may be formed between any two individual values which possess the important property that they remain invariant under transformation (i.e., when re-expressed in different units of measurement). This may be illustrated by calculating the ratio of 10 inches of rain to 5 inches. This ratio (i.e., 2) remains constant if the levels of precipitation are re-measured in millimetres (e.g., 10" = 250 mm, 5" = 125 mm, therefore 250/125 = 2). This invariance contrasts with the interval measure, temperature, in which transformations from one unit of measurement to another alter any ratios which may be formed.

3.4 SCALES AND ANALYSIS

To the purists in the representational school of measurement, the relationships which define each scale influence how it may be used mathematically.

Table 3.1 describes some of the statistical procedures considered appropriate. The clear message presented here is that non-parametric (also termed distribution-free) techniques may be used regardless of scale, but that parametric techniques may only be used with continuous scales. This is a traditional and well-established view which influences a great deal of quantitative work in geography. However, it is not universally accepted, and is frequently violated, often for pragmatic reasons. What costs and benefits are involved in this?

3.4.1 Treating categorical measures as continuous

Categorical data have traditionally tended to be ignored as worthy of quantitative analysis. The following quote illustrates a common reaction. Focusing their attention on nominal measurements, Rhind and Hudson note that:

> little can be done in any statistical analysis of such data apart from counting the frequency of each class, unless the observations are grouped into larger units in which a number of geographical individuals exist.
>
> (Rhind and Hudson 1980: 29–30)

The motivation for this view seems to be that nominal measurements are actually undermeasured continuities – continuous measurements which have been recorded too crudely to illustrate their normal defining relationships. As a result, researchers may argue that they are justified in attempting to analyse them using continuous measurement techniques. Labovitz (1967, 1972), addressing the same sort of problem in sociology, provides similar advice. In this work, it is argued that if the measurements in the data are nominal they ought to be reconceived at least as ordinal, and analysed with continuous data statistics. In particular:

> assumptions, measurement scales, robustness and power efficiency should be thoroughly scrutinised by the researcher before selecting a statistical technique. The evaluation of these factors provides the basis for considering the possibility of balancing the problem of having techniques not quite permitted by the data with the advantages of using a well-developed and clearly interpretable [technique].
>
> (Labovitz 1967: 160)

The logic behind this argument is that continuous measurements provide much greater information on the environmental properties of the data than categorical measurements. Because of this, mathematical and statistical techniques which are developed for continuous measurements are generally more 'powerful' (Blalock 1979). This means that they make better use of the data and can provide more precise information about their structure.

The main cost, however, is that by re-expressing categorical measurements as continuous, the researchers may impose a structure on the data which does not actually exist on the ground (see the discussion on the recording of spatial information in Chapter 2, section 2.5). The problem, therefore, is in achieving a balance between statistical ease and an acceptable analysis of the geographical individual.

3.4.2 Treating continuous measures as categorical

The reverse problem may also arise in which continuous measurements may be treated as though they were categorical. For example, the ratio measure 'distance' may be reduced to a nominal measure by aggregating detailed information into generalised sets such as 'near places' and 'far-away places'. Though this seems to be illogical given the time and effort needed to collect the data in the first place, there may be situations where categorising continuous data is essential to preserve their integrity or the confidentiality of the sources.

Official data released for analysis may undergo categorisation into arbitrary groupings if the recording instrument used to collect them has limited precision and rounds the figures internally, or if geographical 'individuals' can be identified from the collected data. The latter problem occurs frequently, particularly in secondary data collected by government. The triennial Census of Employment illustrates the problem particularly well. The purpose of this Census is to collect information on the occupational and industrial structure of the country. The data are provided by individual firms and are subject to confidentiality restrictions set out in the Statistics of Trade Act (1947). These prohibit the use and dissemination of research results if they allow individual firms to be identified. From the point of view of the firm, these restrictions are vital if strategic information is not to be given away to competitors. The application of the confidentiality rules can apply either to the firms themselves or to the data. Either could be aggregated to a higher level so that the information is presented by industrial classification rather than by firm, or by statistical region rather than by establishment. In either case, the process of classification makes detailed locational studies difficult.

A second reason for categorising continuous measures is to provide a standardised way of recording behavioural information in social surveys or stimulus–response experiments (Miller 1956; Luce et al. 1976). Such standardisation is needed because there are likely to be as many types of behavioural response to a survey question as respondents. Without a structure such as the response sets displayed in Table 3.2, it might prove to be impossible to process, let alone interpret, the results of the survey. The choice of sets, and their number, may be determined from a pilot survey, from existing work or from theory. As with the choice of areal sets used to record and display spatial data, these decisions may confound any interpretations which may be drawn.

3.5 SUMMARY

The Stevens classification is widely used to describe measurement issues in the social and physical worlds. The basis of this approach is the belief in the correspondence between these worlds and the characteristics of number systems. In the forty years since it was first published several modifications have been suggested for it, either by identifying 'ordered metrics' which lie somewhere between the ordinal and the interval (see, for example, Coombs 1950; Thrall *et al.* 1954), or by defining multidimensional classification procedures using more than one dimension simultaneously (Kruskal and Wish 1978). In spite of these developments, and the debates between the purists and the pragmatists, the notion of a series of scales remains popular.

Though few geographers would argue forcefully from the purists' viewpoint, urging instead a degree of experimentation and exploration, much research has been aimed at incorporating procedures and techniques which are scale-related. Thus, there is now considerable agreement that categorical data are best described by techniques developed specifically for them, rather than by assuming that categories are actually under-measured continuities. This acceptance has implications for the sorts of descriptive and analytical procedures which are popularised in class teaching and in research. Some of these are outlined in the next chapter.

4

SUMMARISING GEOGRAPHICAL DATA

4.1 INTRODUCTION

Collecting information is one thing, making use of it is another. Because geographical data sets may be very large, some form of numerical summary is usually required if the key features of the data are to be identified and communicated to other researchers. An effective summary usually involves the description of the following five characteristics of the data:

1 the size of the data set (patterns emerging from small data sets are more likely to be idiosyncratic than those emerging in larger data sets);
2 the 'shape' of the data (to identify 'typical' and extreme values);
3 the central tendency of the data (the tendency for observations to cluster around a typical value or values);
4 the scatter of the observations about this central, typical value;
5 irregular aspects of the data which cannot be accommodated by characteristics 2 to 4.

Such information may provide detailed clues to the underlying form of the full data set, removing the need for an observation-by-observation description. However, before this summary can be provided, the researchers need to ensure that their data are meaningful in their own terms. Essentially this involves some form of quality control to check that zero figures and the letter 'o' have not been confused, that survey responses are recorded in the given range, and that decimal points are in the correct place. These are relatively simple matters which are often overlooked as the researchers rush to compute. (For details of the quality control checks performed in the Cardiff Consumer Survey, see Guy *et al.* 1983.)

This chapter describes some of the measures which can help the geographer summarise a data set. It begins by introducing some of the key ideas of Andrew Ehrenberg's approach to data reduction, introduces numerical and graphical measures to assess central tendency and dispersion, and finishes with a consideration of some of the methods available for assessing association among several variables simultaneously.

46

4.2 PRELIMINARY DATA ANALYSIS

The five steps described above form the core of a preliminary analysis of a geographical data set. Such analysis should always be performed when data are collected in order to gain some impression of their structure. Frequently, such analysis will yield sufficient information to make more sophisticated, model-based analyses unnecessary. It may also identify facets of the data which violate assumptions of these more sophisticated models, a factor which many of the more popular commercial packages still fail to assess adequately (for further details, see Ehrenberg 1975, 1982; Chatfield 1982, 1983; Chatfield and Collins 1980: Chapter 3; Wetherill and Curram 1984).

There are no hard and fast rules governing preliminary data analysis but useful guidelines have been provided by Andrew Ehrenberg in an approach he terms 'data reduction'. This involves simplifying and rearranging the format of the raw data prior to formal analysis in order to make it easier to detect points of similarity and contrast between the observations. The main reason for doing this is to identify key features of the data which must be described and interpreted if the analysis is to be considered satisfactory. It also provides a useful check on the quality of this analysis by highlighting features of the data which may cause problems if ignored.

The following steps provide some indication of how data reduction may be applied in practice. First, the data presentation should be simplified if possible by rounding digits so that quick mental calculations may be made. Second, the simplified raw data should be rewritten so that figures which are to be compared are close together. Third, the order of the variables in the data set should be rewritten to highlight the major differences between the variables. The purpose of these steps is to obtain a quick impression of the character of the data, providing a baseline against which more sophisticated and detailed analyses may be compared.

The practical value of these steps can be illustrated using the data in Table 4.1, which show the absolute (unadjusted) monthly unemployment figures for January to December 1986 in eleven regions of Great Britain. At first glance this table looks complex, appearing to represent a count of regional unemployment down to the last individual. The table contains 132 separate entries (11 rows by 12 columns) and is organised with each month's figures presented as columns, and the regions presented as rows. The rows are arranged in the order used to reference them in NOMIS (i.e., South-East is region 1, East Anglia is region 2, etc.); it does not imply a rank ordering.

The results of applying Ehrenberg's first step to Table 4.1 are displayed in Table 4.2, in which the raw figures have been rounded to the nearest 1,000. The choice of nearest thousand is not arbitrary but corresponds to Ehrenberg's 'two-variable digit rule'. This is a device which simplifies apparently complex figures by excluding all but two variable digits in each

47

Table 4.1 Raw unemployment data for 1986 by region

	Jan.	Feb.	Mar.	Apr.	May	Jun.	Jul.	Aug.	Sep.	Oct.	Nov.	Dec.
SE	398,705	396,757	391,158	385,364	375,717	367,488	374,441	376,338	376,718	366,782	363,941	365,647
EA	87,599	87,888	86,712	85,637	84,144	81,345	82,113	81,792	82,222	80,138	81,039	81,925
LD	413,938	412,897	406,196	409,376	404,297	404,913	411,365	415,149	415,141	403,606	397,104	398,905
SW	220,013	218,002	211,783	208,312	202,958	195,959	199,602	200,811	204,605	201,970	203,800	205,247
WM	356,316	353,955	348,920	349,000	344,233	341,745	346,690	347,844	356,121	343,453	338,437	336,464
EM	209,643	208,207	205,944	205,847	201,945	199,324	202,568	202,508	204,596	198,687	197,737	198,542
YH	324,301	321,312	316,217	320,476	316,773	311,916	315,964	314,343	322,810	311,415	308,805	309,771
NW	463,793	458,228	449,969	454,069	449,166	443,801	450,218	448,038	455,932	438,893	435,638	436,784
N	246,231	242,740	238,909	240,255	236,145	231,926	233,027	230,709	236,357	228,234	228,395	228,319
WL	190,368	188,405	184,247	183,857	179,225	173,708	175,186	173,986	180,370	174,105	173,342	173,546
SC	371,117	367,198	359,318	356,676	351,572	351,359	358,988	358,570	363,037	359,236	360,125	365,217

Codes: SE = South-East; EA = East Anglia; LD = London; SW = South-West; WM = West Midlands; EM = East Midlands; YH = Yorkshire and Humberside; NW = North-West; N = North; WL = Wales; SC = Scotland

Source: NOMIS

reported figure. To see its relevance here, consider the figures in row 1 of Table 4.1, which correspond to unemployment in the South-East region. The first thing to note is that each of these figures contains six digits. By scanning across this row, beginning from the left, it is clear that all the figures begin with the digit 3 – a constant. The second digit of each figure varies a little, ranging from 6 to 9 (4 and 3 occurrences respectively). The third digit varies much more widely – 8, 6, 1, 5, 5, 7, 4, 6, 6, 6, 3, 5 – as do the remaining digits. This same pattern of variability (a constant followed by five variable digits) also applies to the remaining regions with the sole exception of East Anglia, whose unemployment figures contain five rather than six digits. These may be rounded to the nearest hundred if desired.

Table 4.2 Rounded unemployment data for 1986 by region (in thousands)

	Jan.	Feb.	Mar.	Apr.	May	Jun.	Jul.	Aug.	Sep.	Oct.	Nov.	Dec.
SE	399	397	391	385	376	367	374	376	377	367	364	366
EA	88	88	87	86	84	81	82	82	82	80	81	82
LD	414	413	406	409	404	405	411	415	415	404	397	399
SW	220	218	212	208	203	196	200	201	205	202	204	205
WM	356	354	349	349	344	342	347	348	356	343	338	336
EM	210	208	206	206	202	199	203	203	205	199	198	199
YH	324	321	316	320	317	312	316	314	323	311	309	310
NW	464	458	450	454	449	444	450	448	456	439	436	437
N	246	243	239	240	236	231	233	231	236	228	228	228
WL	190	188	184	184	179	174	175	174	180	174	173	174
SC	371	367	359	357	352	351	359	359	363	359	360	365

Source: Table 4.1

Rounding figures in accordance with this rule provides researchers with a way of identifying points of similarity and contrast in the data which may need to be described in detail at a later stage. In this case, by allowing rounding to the nearest hundred rather than thousand, it has highlighted the relative lowness of East Anglian unemployment. Some researchers may, however, be loathe to do such rounding on the grounds that this throws away information and, of necessity, standardises the raw data. Several arguments may be raised against such a view. First, this is a preliminary analysis designed to see general trends and departures from trends; it is not the final analysis. Second, so long as a record is kept of the raw figures no information is actually lost from subsequent analysis. Third, the range of the raw figures – 80,138 (East Anglia, October) to 463,793 (North-West, January) – is affected by less than 1 per cent if they are rounded to the nearest thousand. Most researchers would be hard pressed to interpret this residual information which accounts for so little of the data. Fourth, the precision of the raw figures is artificial in the first place given the nature of the measurement system used (based on a count of monthly unemployment claimants) and the uncertain influence of government job creation schemes which have been applied unevenly throughout the country. In effect, there are sound practical reasons for rounding raw figures in a preliminary analysis, not least of these being the removal of attention from artificially precise raw data.

The second and third of Ehrenberg's steps may now be applied. These involve rewriting the rounded data table to pick out figures which are similar and those which are markedly different (Table 4.3). A useful strategy is to place figures which are roughly similar in columns and arrange the columns in size order, smallest numbers to the left. This further reduces the information which the researchers have to absorb and makes what remains somewhat easier to assimilate.

Having done this, the following features of the data now begin to emerge. First, there is a systematic pattern in the severity of regional unemployment as measured absolutely, with East Anglia being consistently best placed in the regional rankings of Table 4.3 and the North-West worst. Second, the relative position of the regions is constant except for the change in rank order between the East Midlands and the South-West for June, July and August, when unemployment fell faster in the South-West. Third, the pattern of variability within each region over the year is relatively small compared with the absolute levels of unemployment. Only the figure for the South-West varies by more than 10 per cent of the upper limit. Fourth, the pattern of change in absolute unemployment is also similar between the regions (Table 4.4) declining in most regions compared with the preceding month in February to June, rising in July to September, falling in October and November and rising again in December. There are, however, exceptions, and these would need to be accommodated in a subsequent analysis.

Table 4.3 Regional rankings of unemployment data for 1986

	EA	WL	EM	SW	N	YH	WM	SC	SE	LD	NW
Jan.	88	190	210	220	246	324	356	371	399	414	464
Feb.	88	188	208	218	243	321	354	367	397	413	458
Mar.	87	184	206	212	239	316	349	359	391	406	450
Apr.	86	184	206	208	240	320	349	357	385	409	454
May	84	179	202	203	236	317	344	352	376	404	449
Jun.	81	174	199	196	231	312	342	351	367	405	444
Jul.	82	175	203	200	233	316	347	359	374	411	450
Aug.	82	174	203	201	231	314	348	359	376	415	448
Sep.	82	180	205	205	236	323	356	363	377	415	456
Oct.	80	174	199	202	228	311	343	359	367	404	439
Nov.	81	173	198	204	228	309	338	360	364	397	436
Dec.	82	174	199	205	228	310	336	365	366	399	437

Source: Table 4.2

Table 4.4 Pattern of change in unemployment data for 1986

	EA	WL	EM	SW	N	YH	WM	SC	SE	LD	NW
Feb.	0	−2	−2	−2	−3	−3	−2	−4	−2	−1	−6
Mar.	−1	−4	−2	−6	−4	−5	−5	−8	−6	−7	−8
Apr.	−1	0	0	−4	+1	+4	0	−2	−6	+3	+4
May	−2	−1	−4	−5	−4	−3	−5	−5	−9	−5	−5
Jun.	−3	−5	−3	−7	−5	−5	−3	−1	−9	+1	−5
Jul.	+1	+1	+4	+4	+2	+4	+5	+8	+7	+6	+6
Aug.	0	−1	0	+1	−2	−2	+1	0	+2	+4	−2
Sep.	0	+6	+2	+4	+5	+9	+8	+4	+1	0	+8
Oct.	−2	−6	−6	−3	−8	−12	−13	−4	−10	−11	−17
Nov.	+1	−1	−1	−1	0	−2	−7	+1	−3	−7	−3
Dec.	+1	+1	+1	+1	0	+1	−2	+5	+2	+2	+1

Source: Table 4.3

4.3 DESCRIBING THE SHAPE OF DATA

The term 'shape' refers to the patterns which are formed when the observations are displayed graphically. These patterns highlight typical values within the given range of observations, and also make it easier to detect observations which are relatively extreme. Both of these aspects need to be considered if the summary is to be adequate.

Graphics provide an ideal medium for describing shape. A variety of graphical devices exists which may be used with different types of data. The following subsections introduce a number of the more commonly used variants.

4.3.1 Time-series plots

One of the most useful types of graph for describing shape in continuous time-series data, such as presented in Table 4.1, is the time-series plot. This consists of a two-dimensional graph in which observations on a single variable are plotted against time. The time units are plotted on the horizontal axis, and the observations are plotted on the vertical axis. The main use of such a graph is to see if the observations form a discernible pattern over time. Three typical patterns which might be seen are:

1 *Trends* – a general tendency for the values of the observations to rise (or fall) consistently over the time period being plotted.
2 *Cyclical fluctuations* – sequences of rising and falling values which may form a regular pattern.
3 *Irregular fluctuations* – sequences which do not form a clearly defined pattern.

Figure 4.1 illustrates this type of plot using the data in Table 4.1. The vertical axis is scaled in blocks of 50,000 persons to accommodate the range of the readings in Table 4.1.

The most obvious patterns to emerge from these plots are the stability of the rank ordering of the regions and the relatively minor fluctuations in regional monthly unemployment throughout the year, findings which support the tentative conclusions given previously. A degree of clustering also appears to be occurring with East Anglia and the North-West standing out as extreme regions. Four regions (North, South-West, East Midlands and Wales) appear to cluster around a monthly unemployment figure of 200,000–250,000, three more (Yorkshire/Humberside, Scotland and the West Midlands) cluster around 320,000–370,000, and London and the South-East cluster around 400,000–420,000, at least in the early part of the year. The plots appear to trend downwards with monthly unemployment towards the end of the year being generally lower than at the beginning. There is also some evidence for a cyclical pattern as most regions appear to have experienced periods of relatively low unemployment in June and relatively high unemployment in September.

This latter inference cannot be established with certainty from the graph because of the difficulty of reading accurately any of the monthly unemployment figures. This is a problem which affects all types of graphical display and not just time-series plots. It illustrates a major shortcoming with the use of graphs in data analysis which is that graphs are excellent as summary devices only if the intention is to communicate general features of the data, notably qualitative differences. In contrast, they are poor, and frequently useless, for communicating quantitative information. Indeed, if quantitative information is required it is generally more appropriate to use the original data table rather than rely on a graph.

It is also important to remember that the shapes displayed in a graph depend on how it has been drawn. In Figure 4.1, all eleven regions have been included in a single graph. To facilitate this, the vertical axis has been standardised to accommodate the range of unemployment figures within and between the regions. At this scale a clear enough picture emerges of the differences in absolute unemployment between the eleven regions. However, the same scale does not adequately describe the variability of monthly unemployment within specific regions. If attention were to be focused on a specific region, a scale more appropriate to its variability in unemployment should be chosen.

This approach, focusing on East Anglia and the North-West, is illustrated in Figure 4.2. This figure makes it easier to interpret the generalised lines in Figure 4.1 associated with these 'extreme' regions. Both show that unemployment declines over the periods January to March and May to June (by about 7,000 in East Anglia and 20,000 in the North-West), rises in July (1,000 compared with 6,000), falls in August (a few hundred compared with a few thousand), rises in September (a few hundred compared with nearly 10,000), falls in October (2,000 compared with nearly 20,000) and rises in December (about a thousand in both regions).

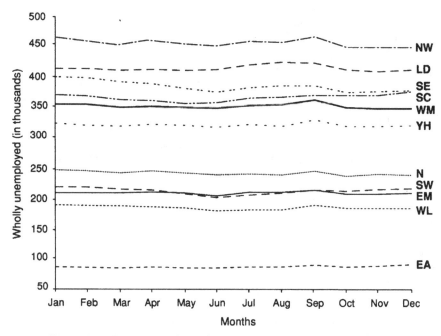

Figure 4.1 Time-series plots of monthly regional unemployment in 1986

The direction of unemployment change diverges in February, April and November. The different scales used in these plots provide better information on the quantitative changes which occurred within each region. They are not, however, most suitable for comparing the magnitude of the changes between the two regions, because similarly-sized oscillations in the two plots represent different absolute levels of change.

This raises an important difficulty: the wish to search for patterns between regions makes it more difficult to search for patterns within regions, and vice versa. This is a major practical problem in analysis because the assessment of the relationship of patterns within a series of observations to those between different series is a fundamental issue in describing many geographical data sets. The preliminary analysis and time-series plots of the data in Table 4.1 have thrown the issue into clear relief.

4.3.2 Histograms

A second graphical device which is useful for describing the shape of continuous data is the histogram. This graph makes use of the fact that consistent patterns in data often only emerge when observations are grouped together into exhaustive, mutually-exclusive categories. It is thus of value if the continuous data may be grouped together into a series of

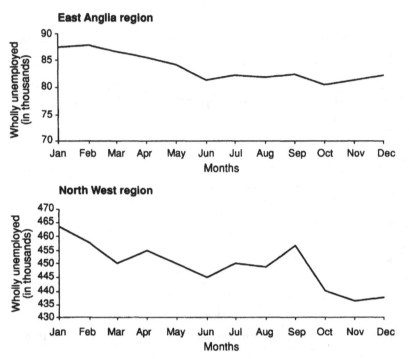

Figure 4.2 Time-series plots of East Anglian and North-Western unemployment in 1986

discrete sets (a user-defined categorisation similar to those described in Chapter 2, section 2.6). If this is possible, the histogram provides information on the frequency of each of the sets in the raw data (i.e., on how often each of them occurs). This information is useful because it provides some insight into typical values (i.e., values which occur frequently) and scatter (i.e., the dispersion of the observations around these typical values).

Histograms of the regional unemployment data in Table 4.1 are presented in Figure 4.3. Each was produced as follows. First, calculate the range of the observations (i.e., the difference in value between the smallest and largest observation in each region). Second, subdivide this range into a small number of equally-sized sets which do not overlap yet cover the whole range. Third, allocate the observations to these sets (alternatively termed 'class intervals'). Fourth, draw rectangles centred on the midpoint of each class interval which are proportional in area to the number of observations contained in it.

The procedure may be illustrated using the East Anglian data. The figures range from 80,138 to 87,888 (or from 80,000 to 88,000 if the figures

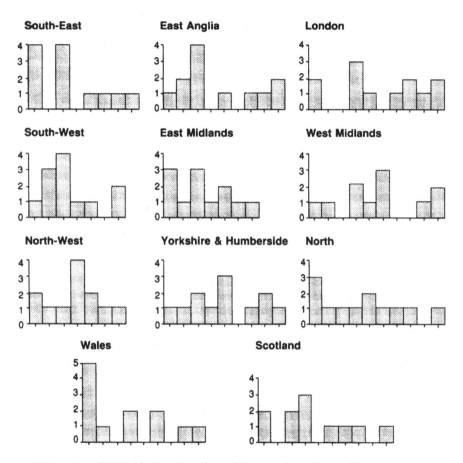

Vertical scale: Relative frequencies of monthly unemployment per class.
Horizontal scale: Unemployment classes centred on mid-points.

	1	2	3	4	5	6	7	8	9	10	11
SE	365	370	375	380	385	390	395	400			
EA	80	81	82	83	84	85	86	87	88		
LD	398	400	402	404	406	408	410	412	414	416	
SW	196	200	204	208	212	216	220				
WM	336	338	340	342	344	346	348	350	352	354	356
EM	198	200	202	204	206	208	210				
YH	308	310	312	314	316	318	320	322	324		
NW	435	440	445	450	455	460	465				
N	228	230	232	234	236	238	240	242	244	246	
WL	174	176	178	180	182	184	186	188	190		
SC	352	354	356	358	360	362	364	366	368	370	372

Note: Histograms produced using MINITAB.

Figure 4.3 Histograms of regional unemployment in 1986

55

are rounded to the nearest thousand). This range may be subdivided into nine sets, each corresponding to 1,000 persons, beginning at 79,500 and finishing at 88,499. Thus set 1 is used to represent the number of months in which the unemployment figure was between 79,500 and 80,499, set 2 those months in which it was between 80,500 and 81,499, and so on. Note that in doing this the time-series information is being ignored. The frequency distribution for East Anglia using these sets is given in Table 4.5. When this information is plotted for display, the frequencies are centred on the midpoints of each set (namely, 80,000, 81,000 etc.).

Table 4.5 Frequency distribution of East Anglian monthly unemployment data for 1986

Set	Range	Frequency
1	79,500–80,499	1 month
2	80,500–81,499	2 months
3	81,500–82,499	4 months
4	82,500–83,499	none
5	83,500–84,499	1 month
6	84,500–85,499	none
7	85,500–86,499	1 month
8	86,500–87,499	1 month
9	87,500–88,499	2 months

Before creating a histogram the researchers need to decide the number and size of the class intervals to be used. These are interrelated decisions as the choice of size clearly influences the number of equally-sized sets which may be obtained within the available range. Many standard computer packages such as GLIM 3.77 and MINITAB produce histograms in response to a given set of commands. These automatically calculate the range of the data and generate the histograms using default configurations programmed into the packages. As a result the graphs produced may not be entirely satisfactory (their defaults can, however, be changed by the user).

To illustrate this, consider the effect on shape of reducing the number of sets to be used with the East Anglian data from nine to three. The resulting histogram is presented in Figure 4.4 along with the original nine-set version. In this modification, each set corresponds to a class interval of 3,000, e.g., set 1 to 79,500–82,499 (midpoint 81,000), set 2 to 82,500–85,499 (midpoint 84,000), and set 3 to 85,500–88,499 (midpoint 87,000). The pattern of the observed frequencies is changed radically as a result even though the numbers being classified and the logic of histogram construction remain the same. In the original version, the shape tends to be skewed to the left (i.e., there are relatively more months with lower unemployment than months with higher unemployment in the data set), though there is evidence for a 'U' shape with the raised frequency of the 87,500–88,499 set. In the modified version, the predominant shape to appear is the 'U' shape rather than the skew.

The reason for this change in shape and interpretation is simply the change in the definition of the sets. It is not possible to state dogmatically that one histogram is better than the other as this decision depends on other factors. However, it should be noted that these histograms are only two out of a large number of possible histograms which could have been drawn. At one extreme, each observation could have been allocated to a unique class to give a histogram with 7,750 sets, most of which would be empty. At the other extreme, all the observations could have been allocated to a single set in which the shape produced would be a single vertical rectangle with no dispersion around it. Neither would be of any descriptive value. The choice of an appropriate histogram is thus a classification problem, identical in form to those described in Chapter 3, and requiring the same sort of approach.

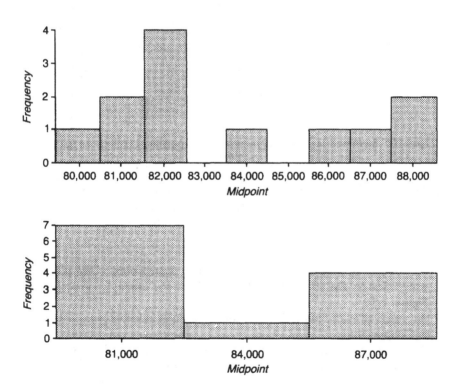

Figure 4.4 Alternative histogram of East Anglian unemployment in 1986

4.3.3 Stem-and-leaf plots

One of the biggest drawbacks with histograms is that they obscure the value of the individual observations. It is simply impossible to determine the individual values of observations in a histogram set because the rectangle used to summarise its frequency is drawn at the midpoint. A graphical device which overcomes this problem but which retains some of the visual advantages of the histogram is the 'stem-and-leaf' plot (Figure 4.5), which is rather like a histogram turned on its side. This is constructed in a similar way to the histogram by calculating the range of the data and subdividing it into intervals of fixed length. Chatfield (1983: 360) notes that these class intervals should be either 0.5, 1 or 2 times a power of 10, a fact which can sometimes be restrictive.

(a) Stem and Leaf of SOUTH-EAST			(b) Stem and Leaf of EAST ANGLIA			(c) Stem and Leaf of LONDON		
Leaf Unit = 1,000 n = 12			Leaf Unit = 100 n = 12			Leaf Unit = 1,000 n = 12		
Number			Number			Number		
1	36	3	1	80	1	1	39	7
4	36	567	5	81	0379	2	39	8
5	37	4	(2)	82	12	2	40	
(3)	37	566	5	83		3	40	3
4	38		5	84	1	5	40	44
4	38	5	4	85	6	6	40	6
3	39	1	3	86	7	6	40	9
2	39	68	2	87	58	5	41	1
						4	41	23
						2	41	55

Figure 4.5 Stem-and-leaf plots of monthly unemployment (selected regions) in 1986

At first sight, stem-and-leaf plots look like a jumble of numbers and digits which are merely separated by a vertical line. However, this 'jumble' is actually organised systematically. Five pieces of information are generally displayed in a stem-and-leaf plot:

1 the number of observations which are to be represented in the plot (this is signified by the letter *N*, and equals 12 in the examples presented in Figure 4.5);
2 the value of each leaf unit (i.e., the value each number presented on a leaf represents);
3 the value of the stem classes, arranged in increasing order (note, these represent the lower boundaries of the class intervals, not the midpoints as in a histogram);
4 the leaf values, arranged in increasing numerical value away from the stem;

5 a range of cumulative frequencies which sum the number of observations in each class interval from either end of the distribution. The middle of the distribution (i.e., the point in the plot where half the numbers of observations are below and half are above) is marked by the bracketed figure.

As before, this type of graph may be illustrated using the East Anglian unemployment data. The stem-and-leaf plot of these data shows that there are twelve items of data classified in the graph ($N=12$), and that the leaf units are to be interpreted as hundreds. To clarify this interpretation, consider the first three lines of the graph. The column headed 'Number' displays the cumulative number of observations being classified in each row and in earlier rows, i.e., one observation only in row 1 and five in rows 1 and 2. The value of the observation in row 1 is given as 801 (80 from the stem, plus 1 from the leaf). To change this into the absolute units of the raw data, all one needs to do is multiply the 801 by the leaf unit (in this case, 100): $801 \times 100 = 80,100$. Each row corresponds to a set. Thus row 1 corresponds to months when unemployment was between 80,000 and 80,999, set 2 months between 81,000 and 81,999, and so on. In the second row, the leaf part displays four observations which, when combined with the stem, are seen to be 810, 813, 817 and 819. By multiplication as before, these become 81,000, 81,300, 81,700 and 81,900. Row 3 is interpreted in the same way: two observations being classified which are 82,100 and 82,200. The only difference between this row and the others is that the number column does not contain the cumulative count. Instead, within brackets, it presents the number of observations classified in the row. The reason for the brackets is to inform the researchers that the midpoint of the data distribution has been reached. Half of the observations in the data are classified between the two end stems and this row. To see this, add 1 to the cumulative counts in rows 2 and 4. This produces the value 6, which is exactly half of the number of observations to be plotted. This information is provided to allow an estimate of the central tendency of the data to be determined directly from the graph (see the discussion on the median in section 4.4).

As far as shape is concerned, stem-and-leaf plots offer essentially the same information as histograms, except that some (rounded) figures may be read off and used in calculations. They are similarly affected by classification problems, which means that care should be exercised in interpreting any patterns they display.

4.3.4 Boxplots

A graph which provides a clearer visual impression of scatter in continuous data is the boxplot, or box-whisker plot (Figure 4.6). This consists of a rectangular box and two 'whiskers' which protrude from either end. To interpret this graph, consider the boxplot of East Anglian unemployment. The ends of the whiskers correspond to the minimum and maximum values in the data. Thus the end of the left-most whisker corresponds to 80,138 (the unemployment figure for October), while the end of the right-most whisker corresponds to 87,888 (the unemployment figure for February). The linear scale printed beneath the plot gives some impression of the magnitude of the observations. The vertical edges of the rectangle correspond to the lower and upper quartiles of the data. The lower quartile is the value below which 25 per cent of the observations lie. The upper quartile is the value above which 25 per cent of the observations lie. The length of the rectangle corresponds to the inter-quartile range, the distance which includes the middle 50 per cent of the observations in the data. Finally, the cross marked in the rectangle corresponds to the median, the measure of central tendency which corresponds to the midpoint of the distribution. However, it is not clear from the graph what numerical values are associated with these measures.

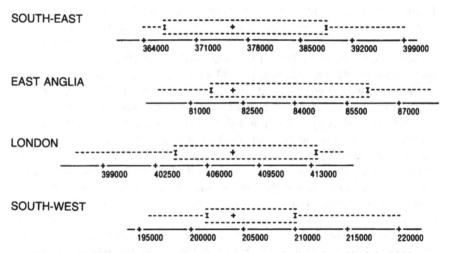

Figure 4.6 Boxplot of monthly unemployment (selected regions) in 1986

One of the most helpful features of boxplots is that they provide an efficient way of comparing dispersion in a series of data sets. To do this, however, the data need to be measured in the same units and be of the same order of magnitude. This means that the boxplots in Figure 4.6 cannot easily be compared without adjustments being made to compensate for their different scales. An adjusted scale is provided in Figure 4.7, the immediate effect of which is to reduce greatly the clarity of the numerical information in each plot. However, comparisons between regions may be made. Dispersion appears to be smallest in East Anglia, the East Midlands

60

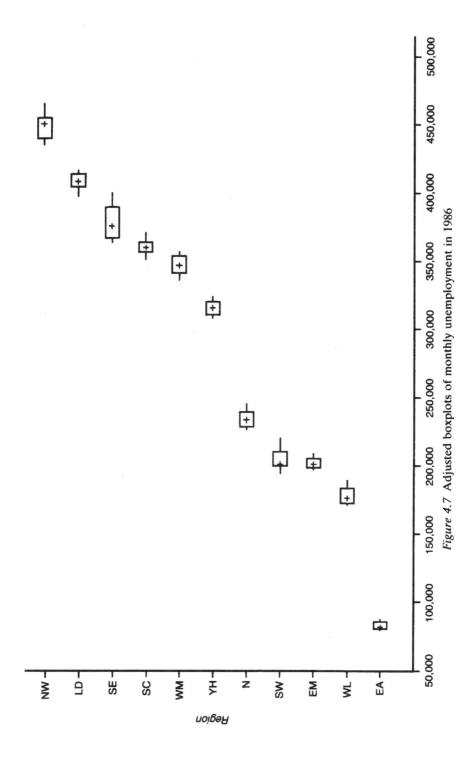

Figure 4.7 Adjusted boxplots of monthly unemployment in 1986

and Scotland, and greatest in the South-East and North-West. Most regions exhibit an extended right whisker indicating that about 25 per cent of their monthly readings are relatively high. In contrast, only London and the West Midlands appear to exhibit the opposite tendency. The scatter pattern for Scotland suggests that it has a roughly equal number of relatively high and low months. It is also interesting to note that the regions appear to form a rough line rising from the bottom left of the plot. If this is a real effect rather than an artefact then a simple linear relationship may be all that is needed to describe the unemployment data. The role of such relationships is discussed in section 4.5 and throughout the remainder of this book.

4.3.5 Cumulative frequency graphs

Boxplots provide some information on scatter, but this is limited to a comparison of medians and the inter-quartile ranges of a series of observations. Frequently, it is more useful to know how many of the observations lie within other ranges, or below or above a given threshold. This information may be calculated from a stem-and-leaf plot if the data set is small, but a more effective graph for larger data sets is the cumulative frequency graph. Figure 4.8 shows cumulative frequency graphs for some tenure data collected from 259 Bristol council tenants as part of a study of the Bristol public housing market (O'Brien 1982). The figure contains two plots, one for cumulative frequencies, the other for cumulative percentages. The data being plotted are in columns 3 and 5 of Table 4.6.

Cumulative frequencies are calculated directly from the observed frequencies (i.e., the second column of data). The figure in row 2 is added to the figure in row 1 to give the total number of tenants who have lived for less than two years in their current tenancy, and so on. These

Table 4.6 Cumulative frequency data: duration of previous tenancy among Bristol council tenants

Tenancy in years	Number	Cumulative frequency	Relative frequency	Cumulative percentage
<1	13	13	5	5
1–2	27	40	10	15
2–3	48	88	19	34
3–4	37	125	14	48
4–5	22	147	9	57
5–6	14	161	5	62
6–7	24	185	9	71
7–8	19	204	8	79
8–9	0	204	0	79
>9	55	259	21	100

Source: O'Brien (1982)

totals are cumulative frequencies and are displayed in column 3 of the table.

A second way of expressing the raw frequencies, without actually altering the shape of the frequency curve, is to convert them into relative frequencies. This is done by dividing each observed frequency by 259, the number of tenants in the data set. This information gives the proportion of the total associated with each row category. As percentages tend to be easier to understand than proportions, it is usual to express the relative frequencies as percentages. The first row of column 4 in Table 4.6 thus shows that 5 per cent of tenants had lived in their previous dwelling less than one year, whereas 10 per cent had lived there for between one and two years. Column 5 displays these relative frequencies as cumulative percentages.

A quick visual inspection of the two plots reveals that the graphs and horizontal axes are identical, the only difference is the units of measurement used on the vertical axes. This suggests that either measure could be used to convey information about the cumulative structure of the data. Though this is true, the cumulative frequency graphs are to be preferred

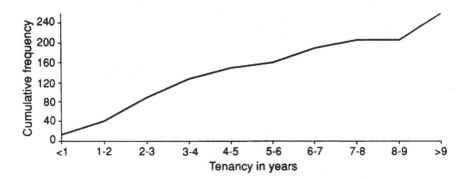

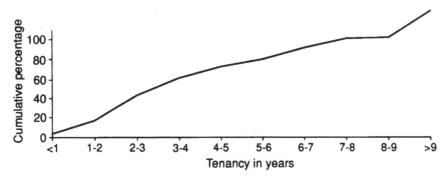

Figure 4.8 Cumulative frequency plots of tenure data

63

because the real data are being graphed, rather than a transformed surrogate for them (the cumulative percentages are essentially cumulative rates per hundred tenants). Moreover, there is sufficient information in the cumulative frequency graphs to produce cumulative percentages if this were desired. Without knowledge of the size of the data set, which may not always be available if the data are secondary, the reverse process cannot be performed.

4.3.6 Barcharts

The graphs presented so far are valuable for displaying the shape of continuously distributed information by categorising it to produce histograms and stem-and-leaf plots. However, graphics designed specifically for discrete data are also available.

Figure 4.9 displays the absolute levels of ethnic unemployment in Great Britain in August 1982, taken from a count of unemployed registrants at job centres. This information is presented for persons born abroad, and persons whose parents were born abroad. The summary graph is termed a barchart. This shows that Asian unemployment was most severe (at about 70,000 persons nationally), followed by West Indian unemployment, African unemployment and 'other' ethnic unemployment. The order of the ethnic categories is arbitrary because they are not sets in a continuum. However, they could be arranged in ascending or descending numerical order, or alphabetical order, if that was considered desirable.

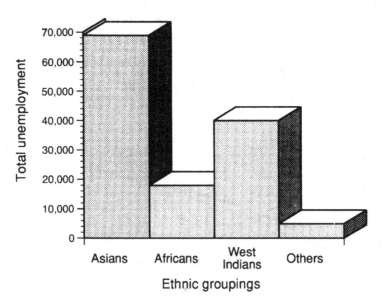

Figure 4.9 Barchart of ethnic unemployment in Great Britain in August 1982

64

4.3.7 Pie-charts

An alternative graphical display which may be used effectively with discrete data is the pie-chart (Figure 4.10). Each segment of the pie corresponds to the absolute level of unemployment associated with each ethnic group. The size of these segments is determined by dividing the group sizes by 131,701 (the overall total), and then multiplying by 360. This produces a result in degrees. These are: Africans 49, West Indians 108, Asians 189 and others 14. Unfortunately, though the quantitative information is needed to calculate the size of the segments, it is invariably lost when the pie-chart is displayed. A quick look at Figure 4.10 shows that Asian unemployment is larger than African unemployment in terms of segment size, but it is not at all clear by how much.

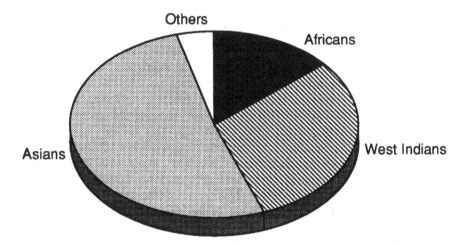

Figure 4.10 Pie-chart of ethnic unemployment in Great Britain in August 1982

4.3.8 Summary

This section has described some of the graphical methods which may be used with different types of geographical data. There is a considerable variety available and they are usually included as standard options in commercial computer packages. However, as we have seen, 'shape' is not an unambiguous concept, but reflects the classification decisions made by the researchers or by the default options programmed into packages. The latter may not be entirely suitable and so should not be accepted at face value. While graphics are useful tools for portraying qualitative information (for example, highlighting the higher unemployment in the beginning of 1986 which was actually measured using a different system from the remainder of the year), they are usually quite poor at conveying quantitative information, for example, on central tendency and scatter. The next section considers this issue.

4.4 SUMMARISING CENTRAL TENDENCY

One piece of quantitative information which is essential to a summary of geographical data is the central tendency of the observations, i.e., the tendency of the observations to group around one or more typical (or representative) values. There are several ways of calculating these values. Three measures which are particularly important are: the mode or modal class, the median or midpoint, and the mean or average value. Versions of these measures are available to describe central tendency in categorised (grouped) and continuous (ungrouped) data.

4.4.1 The mode

The mode, or modal class, corresponds to the most frequently occurring value, or class, in the data set. If calculated on ungrouped data, it refers to the most common individual value in the data set. However, if applied to grouped data, it refers to the most common category. To calculate it in the former case, the data need to be arranged in numerical order and the frequency distribution of each value calculated. This is easily done if the data set is small. For example, in the following sequence of observations (the rounded East Anglian data from Table 4.3) – 80 81 81 82 82 82 82 84 86 87 88 88 – the value 82 (representing a monthly unemployment of 82,000) occurs more than any other. It is therefore the modal value for the sequence. Clearly, if the data set were larger, this procedure might prove to be very tedious.

The calculation is made significantly easier for larger data sets if the data are grouped and displayed graphically in a histogram, barchart or stem-and-leaf plot, because it is obvious from these which class occurs most frequently. The modal classes for the monthly unemployment data may be seen at a glance from Table 4.7. In nine of the eleven regions the modal class corresponds to a single class interval. However, in the South-East and East Midlands there are ties. In the East Midlands, the two classes 197,000–199,999 and 201,000–202,999 occur equally frequently, and in the South-East there is a similar tie between class 362,500–367,499 and class 372,500–377,499. Such distributions are termed bimodal, because there are two modal classes in the data. (The term 'multi-modal' is used to refer to distributions which possess more than two modal classes.)

It is important to note that the precise value of a modal class depends on the size and number of classes used to group the data. The significance of this factor is clearly illustrated by Figures 4.3, 4.4 and 4.5, in which the data are displayed differently. In these figures, the use of different

Table 4.7 Modal classes

Region	Modal class	Number
South-East	Bimodal	
East Anglia	81,500–82,499	4
London	403,000–404,999	3
South-West	202,000–205,999	4
West Midlands	347,000–348,999	3
East Midlands	Bimodal	
Yorks/Humberside	315,000–316,999	3
North-West	447,500–452,499	4
North	227,000–228,999	3
Wales	173,000–174,999	5
Scotland	359,000–360,999	3

category definitions means that the modal classes which may be observed are different. Such extreme dependence on the data classification means that the mode is of very little value as a summary measure other than as a quick 'rule of thumb'.

4.4.2 The median

The second measure used to summarise central tendency is the median. This describes the midpoint of the data distribution: the point at which 50 per cent of the observations lie below it in value, and 50 per cent lie above it. This measure has already been introduced in the preceding discussions on stem-and-leaf plots and boxplots which include it as part of their graphical display. To calculate the median the observations must be counted to determine their number, arranged in numerical order (tedious if the data set is large), and counted off from one end until half the observations have been reached. To illustrate the calculation of the median, consider the data in Table 4.8 which are the unemployment figures for East Anglia for 1986 arranged in ascending numerical order.

Because there is an even number of observations in both sequences, the median cannot be observed directly. Instead, it has to be estimated (the term usually used for this process is 'interpolated'). Logically, because the median represents a half-way point in the data, it should lie somewhere between the sixth and seventh observations in the sequences. A reasonable guess suggests it should be at an hypothetical observation 6.5. To calculate it therefore, the difference between observations 6 and 7 should be divided and half added to the value of observation 6, or half subtracted from observation 7, i.e., 82,168. The medians for the remaining regions are calculated in a similar manner.

The calculation of the median is relatively easy once the middle values have been identified, but is slightly more tricky if these middle values are identical, as is the case with the rounded data for East Anglia. One

Table 4.8 East Anglian unemployment
data for 1986 arranged in ascending
order

Raw figures	Rounded figures
80,138	80
81,039	81
81,345	81
81,792	82
81,925	82
82,113	82
82,222	82
84,144	84
85,637	86
86,712	87
87,599	88
87,888	88

solution to this problem is to treat the median as 82, as in the table, because 50 per cent of the observations have value 82 or less, and 50 per cent have value 82 or more. An alternative, described by Cohen and Holliday (1982: 27), makes use of the so-called 'limits of numbers'. These 'limits' reflect the ranges used to round continuous values to whole numbers.

The number 3, for example, may be used to represent values in the range 2.5 to 3.49, as values between 2.5 and 3 may be rounded up and values between 3 and 3.49 rounded down. To apply this to the calculation of the median the four values of 82 need to be replaced by values within the limits of numbers covered by 82, which are 81.5 to 82.49. As there are four observations of value 82, this limit must be divided into quarters to produce a new sequence of numbers which may be used instead. In other words, replacing the four values of 82 by four ranges: 81.5–81.74 for the first 82, 81.75–81.99 for the second, 82–82.24 for the third, and 82.25–82.49 for the fourth. The modified sequence for East Anglia now becomes: 80, 81, 81, 81.5, 81.75, 82, 82.25, 82.49, 84, 86, 87, 88, and 88. The median is easily calculated as the modified list now contains an odd number of observations. The median is at position 7, i.e., 82.25. As these rounded figures actually represent thousands, the value of the median is really 82,250. This is similar to the figure calculated from the raw data.

The median may also be calculated for grouped data such as the histogram classes in Table 4.5. The procedure used here is as follows. First, divide the number of observations by two to find the 50 per cent point in the distribution (i.e., the number corresponding to the median). Second, starting from the lower end of the frequency distribution sum the number of observations until the set containing the 50 per cent point is reached. Third, subtract the number of observations calculated at stage 2 from the number corresponding to the median (this is used to calculate a scaling factor). Fourth, multiply the scaling factor by the class interval and add to the upper limit of the set immediately below that containing the median. This value is the best estimate of the median from grouped data.

As before, this procedure may be illustrated using the data for East Anglia (Table 4.5). The median point is estimated for the data as follows: the median observation is calculated from step 1 as observation 6, which lies in the third lowest class in the frequency distribution. The number of observations in classes below that containing the median is 3 (1+2, in classes 79,500–80,499 and 80,500–81,499). This gives the value for step 2. The scaling factor in step 3 is calculated by subtracting the figure at step 2 from that at step 1, dividing the result by the frequency of the median class, and multiplying the ratio by the size of the class interval. That is:

1 subtract 3 (step 2) from 6 (step 1) to give 3,
2 divide by 4, the frequency of the class containing the median, to produce a ratio of 3/4,
3 multiply 3/4 by the class interval (1,000) to give 750.

This value is then added to the upper limit of the class below that containing the median, providing an estimate of the median of 81,499 + 750 = 82,249. This compares with the figure estimated from the ungrouped data of 82,167 and the rounded data of 82,250.

4.4.3 The mean

The third and final measure of central tendency to be presented here is the mean or arithmetic average. This measure is defined as the sum of the individual observations divided by the number of observations in the data set. As before, this may be illustrated using the twelve monthly unemployment figures for East Anglia. The sum of these twelve observations is

Table 4.9 Mean monthly
unemployment (1986)

Region	Mean
South-East	378,255
East Anglia	83,546
London	407,740
South-West	206,088
West Midlands	346,931
East Midlands	202,962
Yorks/Humberside	316,175
North-West	448,711
North	235,104
Wales	179,195
Scotland	360,201

1,002,554. Dividing this figure by 12 produces the mean value of 83,546. The mean values for the remaining regions are listed in Table 4.9.

The procedure for grouped data is rather more complicated. It is calculated by multiplying the number of observations in each set or group by a typical value for that group (usually the midpoint), summing these values together, and dividing the total by the total number of observations. For the East Anglian data in Table 4.5 this involves multiplying the midpoint of set 1 (80,000) by 1, 81,000 by 2, 82,000 by 4 and so on, and adding the products to form a grand total (1,003,000). This is then divided by 12 to produce an estimate of the mean of 83,583.

Unlike the other measures of central tendency, the mean makes use of all the information contained in the data rather than just the most frequent value or class, or the 50 per cent value. This makes it generally more useful as a method of summarising the characteristics of data sets than either the median or the mode. It also possesses other advantages. First, the data do not need to be arranged in numerical order as any order may be processed. This makes its calculation easier if the data set is large. Second, the mean provides an effective method of summarising a data set by focusing attention on the individual data values, allowing those which are similar and dissimilar to be observed directly. Third, there is a clear relationship between the mean value, the total for the data set and the number of observations, which does not exist for the mode or the median. So long as any two of these are known, it is possible to calculate the third. This is because:

$N \times$ the mean $\quad = $ the total
the total/the mean $\quad = N$
the total/N $\quad = $ the mean

These relationships can be particularly helpful if observations from different data sets are to be combined or if the data are to be used in subsequent analyses.

Consider the following two sequences of data which are of different size, one containing six observations, the other nine. Sequence 1: 68, 24, 43, 61, 23 and 16; sequence 2: 10, 15, 19, 26, 11, 14, 21, 28 and 32. The mean of the first sequence is 39.17, and the second sequence, 19.56. If the two sequences were to be combined a new mean would have to be calculated. One way of calculating it is to add the two means together and divide by 2, producing a rough mean of 29.37. This, however, is different from the correct mean for the data (27.4) which would be produced if all the observations were added together and their total divided by 15. The reason for the inaccuracy is that no attempt is made in the rough calculation to account for the different numbers of observations in the two sequences. A better approach, which yields the correct mean but which does not involve the researchers having to add the original observations, is to calculate a 'weighted mean'.

The steps to be used in calculating this are as follows. First, calculate the total number of observations in the combined sequence (in this case fifteen). Second, divide the number of observations in both sequences by the combined total to produce two scaling factors (e.g., 6/15 = 0.4 and 9/15 = 0.6). Third, multiply the two sequence means by the appropriate scaling factor and add together, i.e., (39.17 × 0.4) + (19.56 × 0.6). This produces the correct mean value of 27.4 for the combined data. This flexibility is of great value to geographers who frequently collect data in distinct sequences which then need to be combined.

4.4.4 Robust measures

The fact that the mean accommodates all the observations in a data set is considered by some authors to be a disadvantage of the measure. This is because the mean is sensitive to extreme values and can provide a measure of central tendency which is not typical of the bulk of the observations.

To illustrate this consider the following five numbers: 1, 2, 3, 4, and 5. The total of these is 15 and the mean is 3. This value lies exactly half-way in the data distribution, i.e., it is equivalent to the median value. Two observations lie below the mean, 1 and 2, and two lie above it, 4 and 5. The differences in value between the mean and the upper and lower outlying values are also equivalent. For this sort of data distribution therefore, the mean appears to be an adequate description of central tendency. However, if just one of the figures were changed, for example if the value 5 were mis-typed as 15, or if it corresponded to some freak phenomenon, the total and mean would change dramatically. The total would rise to 25 and the mean to 5. Four observations would lie below the mean with a range of 4 units, whereas only one value would lie above it, with a range of 20. The mean would no longer be typical of any of the individual observations whereas the median would have been unaffected.

Given this tendency for the mean to be pulled towards extreme values and away from the main body of observations (as is very likely in geographical hazard studies) some researchers argue that a less sensitive, more robust measure, such as the median, should be calculated in preference. This is one of the key arguments put forward in 'exploratory data analysis', an approach to data analysis proposed by Tukey (1977), and illustrated in geography by Cox and Anderson (1978), Cox and Jones (1981), Jones (1981, 1984) and Burn and Fox (1986).

Robust measures are measures which are resistant to extreme values. The median is resistant in the sense that it ignores data values altogether, except in so far as they are used to rank the data in size order. This means that the effect of extreme values on the measure of central tendency is reduced. However, for many research purposes the median is a cumbersome device to use (see the previous discussion), so a number of modified strategies have been devised to make the mean more robust. One strategy is to order the data by size and exclude a small number of observations from either end. Having done this, the usual formula for the mean is applied to the remaining data. To illustrate this process, which is termed 'trimming', consider the second sequence of five numbers. If one observation were removed from either end the sequence would be: 2, 3, and 4, which has a median of 3 and a trimmed mean of 3 (i.e., 9/3 =3). If two observations were removed from either end, the sequence would contain only a single value, 3. In so far as it makes sense to calculate medians and means for a single number, the median and trimmed mean are 3.

A second strategy again involves ordering the data by size. Having done this, the two most extreme values are replaced by the next most extreme pair in the sequence. Thus, the sequence: 1, 2, 3, 4, and 15, would be replaced by 2, 2, 3, 4 and 4. The median of this reformed sequence is once again 3, as is the mean. The process may be repeated a second time, replacing the extreme pair of values by the second most extreme pair. In this case this means replacing the 1 and 15 by the 3. The reformed sequence is now: 3, 3, 3, 3 and 3, of which the median and mean values are 3. This process is known as Winsorising. Clearly, given this contrived example, the median and the two robust means give more appropriate information about the underlying data than the unmodified mean.

Does this provide an improved description of central tendency for the data? In the example given above the answer is clearly 'yes' because the median and robust means stabilise at a constant value which can be regarded as reasonably typical of the individual observations. To see if this is a general rule consider the following skew data: 0, 0, 0, 0, 0, 2, 2, 3, 15 and 40. The summary statistics appropriate to this sequence are: standard mean – 6.2; trimmed mean – 1.0; Winsorised mean – 1.0, and median –

Table 4.10 Robust measures of mean monthly
unemployment for 1986

Region	Mean	Trimmed mean
South-East	378,255	377,641
East Anglia	83,546	83,453
London	407,740	408,063
South-West	206,088	205,709
West Midlands	346,931	347,040
East Midlands	202,962	202,817
Yorks/Humbs	316,175	316,100
North-West	448,711	448,510
North	235,104	234,678
Wales	179,195	178,663
Scotland	360,201	359,994

1.0. Once again, there is agreement between the robust measures and considerable disagreement with the standard mean, which has been drawn towards the upper end of the distribution by two extreme values.

These examples using fictitious data show that the robust measures can give a substantially different impression of central tendency than the standard mean. However, Table 4.10 shows that there is very little real difference in value between the various measures when applied to the data in Tables 4.1 and 4.3. Indeed, most of the measures differ by less than 0.2 per cent, a trifling figure compared with the 620 per cent reduction achieved in the contrived example given above.

4.4.5 Which to use?

The choice of which measure to use to represent central tendency depends on why such a measure is required. The mode, for example, is most useful whenever the research requires a quick, and relatively informal, indication of a typical data value. The median is most useful if a measure of the midpoint of the data distribution is required. The mean is most useful if knowledge of the exact values of the observations is important. In each of these situations, the information needs of the research differ, ranging from a qualitative impression to detailed quantitative information. If the latter is not required, then the mode or median might provide sufficient information on central tendency. However, if the data are to be used in comparative analyses where data sets are being compared or used for inference (see Chapter 6), then more detailed quantitative information is a requirement. Because of its tractability, the mean is generally preferred to the median in these situations.

As was shown in the previous examples, the perceived advantages of the robust means over the standard mean depend on the data set. However, there are two major deficiencies with their use. First, they emphasise central tendency to the detriment of extreme values, when detailed knowledge of the latter may be crucial. Second, they obscure the calculation of the summary measure, thus making it rather more difficult to relate the measure to the raw data values.

The first of these is particularly important in studies of 'hazards' – such as famine, extreme floods, high tides, global changes in sea level – where knowledge of the frequency and size of the extreme points is crucial. In such contexts, the mode and median do not provide any information on the extremities of the data, while the robust measures are actually designed to reduce their affect on the revised mean. Whilst the resulting number is *more typical of the data, it excludes a key aspect of the dynamism of the* processes generating the data in the first place, and may lead to insufficient attention being paid to the frequency and magnitude of the extreme points. The second deficiency arises because the calculations involved in generating robust measures make it difficult to relate the measure back to the raw data. The most significant problem is that the measure is not immediately interpretable in terms of the raw data. This can lead to confusion, and make additional use of the raw data in comparative analyses difficult.

Whilst it is clear that the standard mean is sensitive to extreme data values and so can give a misleading impression of central tendency, it can be argued that this is actually an advantage, as it requires researchers to relate the measure to the raw data. Such a process is crucial if the data are to be described properly. Because of its sensitivity, the mean can identify both the number and magnitude of the extreme points and provide researchers with suitable analytical prompts. For example, if a data point is extreme, the researchers must consider why this is the case and develop some appropriate strategy to handle it. This might involve removing it from the raw data and assigning it to a second data set which is analysed separately. By focusing on the single measure which 'accommodates' extreme points, it is potentially possible for these questions to be disregarded or undervalued.

4.5 SUMMARISING SCATTER

Measures of central tendency provide a way of obtaining useful quantitative information on typical and atypical values in a data set, but on their own, they do not provide sufficient information to describe it fully. In addition, some quantitative measure of the scatter or dispersion of the data is required. Much of this information is already provided graphically in histograms, stem-and-leaf plots, barcharts and boxplots, but rarely in an easily digestible form, and hardly ever in a form which may be used in subsequent analysis. This is only to be expected: graphs are poor at communicating quantitative information. To provide this sort of information, a variety of numerical summary measures should be used instead. These include: the range; the inter-quartile range; percentile ranges; the mean absolute deviation; the variance and standard deviation; and the coefficient of variation. Some of these have already been introduced in the discussion of graphical displays.

4.5.1 The range

One of the simplest measures of scatter available is the range. This describes the difference in value between the largest and smallest observations in the data set. For the East Anglian unemployment data the range is 7,750, the difference between 87,888 and 80,138. The main point in calculating the range is that it allows the finite nature of the data set to be displayed explicitly. This reflects the fact that the observations are restricted to a limited range of numerical values, and any patterns or relationships which are identified may be partly by-products of this restriction. It is therefore not sensible for researchers to attempt to generalise these patterns or relationships beyond the numerical limits of the observed data, as both may alter significantly in a different numerical context.

However, in spite of this advantage, the range possesses several disadvantages. First, it only makes use of information on the extreme values in the data set. No account is taken of the size of the data set, or of other observations. These are potentially major disadvantages as there is a general tendency for the range to increase as the number of observations gets larger. Second, the value of the range is sensitive to the corrupting influences of outliers. This means that if there are unusually large or small observations at the extremes of the data the range will be over-extended by them. Third, it is often laborious to calculate as the data have to be ordered or otherwise compared observation by observation. This is an expensive operation even if done on a computer, given that the vast majority of the information being processed is subsequently ignored. Finally, knowledge of the range provides little information which may be used in subsequent analyses.

4.5.2 The inter-quartile range

A second numerical measure of scatter is the inter-quartile range. This has been discussed previously in connection with boxplots, which graph it explicitly. The inter-quartile range describes the scatter of the central 50 per cent of the observations. Its lower boundary is described by the lower quartile, the value below which 25 per cent of the observations occur. Similarly, its upper boundary is described by the upper quartile, the value above which 25 per cent of the observations occur. To calculate this, the data need to be organised in ascending numerical order and counted off until 25 per cent and 75 per cent of the observations have been identified. This procedure is similar to that used to calculate the median.

Because of its similarity to the median (each quartile is essentially the median of the bottom and top half of the distribution) it is usual to report quartiles as measures of dispersion when medians are used to measure central tendency. A derived measure, termed the quartile deviation (Q), may also be calculated to summarise dispersion around the median. Its usefulness is, however, mainly limited to summarising scatter in data which are distributed according to the so-called Normal probability distribution, the importance of which will be discussed in Chapter 5.

4.5.3 Percentile ranges

A natural extension of the inter-quartile range is provided by percentile ranges. These have also been illustrated previously in connection with the cumulative frequency curves. Percentiles are measures which represent the data distribution in terms of percentage units (see the column headed 'cumulative percentages' in Table 4.6). The range corresponds to the difference between the 0th and 100th percentiles, and the inter-quartile range to the difference between the 25th and 75th percentiles. Clearly, any range between two percentiles may be reported. While this provides more information than either the range or inter-quartile range, it is still limited to a comparison of two values at a time. Percentiles are also more complicated to calculate as the data need to be arranged as cumulative frequencies and then transformed into cumulative percentages. As before, if there is a lot of data to process, this preliminary data handling may be tedious and time-consuming.

4.5.4 The mean absolute deviation

The three measures described so far provide useful indicators of scatter between pairs of extreme points. None of them refers explicitly to central tendency or to the remaining data items, though the quartiles may be used in this way. The fourth measure to be described overcomes these drawbacks. The mean absolute deviation measure (MAD for short) is a summary of the absolute differences in value between the mean of the data and the individual data items. The term 'absolute' means that all negative signs are ignored and are treated as though they were positive. It is calculated by subtracting the standard mean from each data item to produce an absolute deviation, summing these values together to produce a total absolute deviation, and dividing this total by the number of observations in the data set.

The mean deviation of the East Anglian data is calculated as follows. First, the mean value (83,546 or 83.58) is subtracted from each individual data value to produce deviations from the mean. Second, the twelve deviations are summed together ignoring the negative signs to produce a total absolute deviation. Third, this total is divided by the number of observations in the data set (12) to produce the mean absolute deviation value. These steps are summarised in Table 4.11. From this table, the MAD is calculated by dividing the total absolute deviations by the number of observations: 28,498/12 = 2,374.83.

The interpretation of this measure is quite straightforward: it states that the observed readings of monthly unemployment in East Anglia lie on average within 2,375 units of their mean of 83,546 (2.51 units of 83.58 if the rounded data are used instead). By referring back to the raw data in Table 4.1, it is clear that five of the monthly readings lie between 81,171 (the lower limit) and 83,546, and two other readings lie between 83,546 and 85,921 (the upper limit). The figures for January to March are conspicuously high compared with the average, the figures for October

Table 4.11 Calculation of the mean
absolute deviation for the East
Anglian data for 1986

Raw data	Difference
87,599	4,053
87,888	4,342
86,712	3,166
85,637	2,091
84,144	598
81,345	2,201
82,113	1,433
81,792	1,754
82,222	1,324
80,138	3,408
81,039	2,507
81,925	1,621
Total	28,498

and November conspicuously low. These months may be outliers, i.e., months in which the figures for unemployment are extreme when compared with the remainder of the year. One of the main uses of a measure of scatter is to identify these extreme values.

4.5.5 The variance and standard deviation

The variance and standard deviation are calculated in a similar way to the mean absolute deviation. The variance is defined as the average of the squared deviations from the mean, the standard deviation as the square root of the variance. To calculate the variance the mean is first subtracted from each of the data items to produce deviations. These are then squared to remove the effect of the negative deviations, summed together to produce the total sum of squared deviations in the data, and then this total is divided by the number of observations in the data set (Table 4.12).

Like the MAD statistic, both of these measures summarise the scatter of the individual readings in the data about their mean value. They differ in that the variance expresses this scatter in terms of squared units, whereas the standard deviation expresses this in terms of the original units of measurement. For the East Anglian data, the variance indicates that, on average, the monthly unemployment values are 6,857,313 squared deviation units from their mean. The standard deviation suggests that on average monthly unemployment values are about 2,619 persons from the mean.

There is a slight complication to note in the calculation of the variance and standard deviation. The standard method of calculating them involves dividing the sum of squared deviations by the number of observations in

77

Table 4.12 Calculation of variance and standard
deviation for East Anglia for 1986

Raw data	Difference	Squared difference
87,599	4,053	16,426,809
87,888	4,342	18,852,964
86,712	3,166	10,023,556
85,637	2,091	4,372,281
84,144	598	357,604
81,345	−2,201	4,844,401
82,113	−1,433	2,053,489
81,792	−1,754	3,076,516
82,222	−1,324	1,752,976
80,138	−3,408	11,614,464
81,039	−2,507	6,285,049
81,925	−1,621	2,627,641

Notes:
Total squared difference = 82,287,750
Variance = 82,287,750/12 = 6,857,313
or   82,287,750/11 = 7,480,705
SD   = Square root of 6,857,313 = 2,619
or   Square root of 7,480,705 = 2,735

'Difference' refers to the difference between the individual raw
data value and the mean value.

Table 4.13 Calculation of variance and standard deviation from
grouped data for 1986

Raw data	N	D	D^2	ND^2
80,000	1	−3,583	12,837,889	12,837,889
81,000	2	−2,583	6,671,889	13,343,778
82,000	4	−1,583	2,505,889	10,023,556
83,000	0	0		
84,000	1	417	173,889	173,889
85,000	0	0		
86,000	1	2,417	5,841,889	5,841,889
87,000	1	3,417	11,675,889	11,675,889
88,000	2	4,417	19,509,889	39,019,778

Notes: N   frequency of each class in data set,
         D   difference between raw data and mean,
         D^2 squared difference,
         ND^2 product of squared difference and frequency.

Mean = 83,583
Sum (ND^2) = 92,916,668
Variance = Total/11 = 8,446,970
SD = square root of (total/11) = 2,906

the data set. However, for theoretical reasons which are to do with
sampling theory, and which will not be considered until Chapter 6, it is
preferable to divide the sum of squared deviations by one less than the
total number of observations (i.e., by $N-1$ instead of N). This makes little

or no difference in data sets which contain 20 or more observations, but may become progressively more important as the data set becomes smaller. It is standard practice for most commercial computer packages to report these modified variances and standard deviations.

Both measures may be calculated for grouped data. The procedure is similar to that described earlier to calculate the mean from grouped data. First, the mean value of the grouped data is calculated. Second, this value is subtracted from the midpoint (or other typical value) of the group, and the difference squared. Third, these squared differences are multiplied by the group frequency and summed to produce an overall total. Fourth, this total is divided by the number of observations in the data set (or $N-1$) to yield an estimate of the variance. The square root of this estimate is the standard deviation. The calculations required are set out in Table 4.13.

4.5.6 The coefficient of variation

The last measure of scatter to be presented here is the coefficient of variation (CV). This measure offers an alternative way of describing the information contained in the standard deviation. It does this by expressing the standard deviation as a percentage of its mean. First, calculate both the mean and standard deviation. Second, divide the standard deviation by the mean value. Third, multiply the quotient by 100. This value is the coefficient of variation. For the rounded East Anglian data, the CV is 3.5 per cent (Table 4.14), which means that the monthly readings are on average within 3.5 per cent of the rounded mean value of 84,000.

The value of the CV measure is that it allows comparisons to be made between data sets whose scatter is measured in different units or, like the regional data, have different orders of magnitude. The CVs of the remaining regions are also displayed in Table 4.14. These show that the average scatter of readings around the eleven means is relatively small, ranging from about 1.5 per cent to 3.5 per cent – a very similar level of performance in spite of the major differences between the regions.

The final column in Table 4.14 lists those months which lie outside limits suggested by the coefficient of variation. A distinct pattern seems to be visible here as the months January–March appear to have had relatively high levels of unemployment in most regions, and October and November relatively low levels. These may be outliers or months in which unemployment in most regions was relatively extreme, and so need further investigation. Such an investigation would show that the measurement system used for January and February was different from the rest of the year and tended to inflate the unemployment figures for all regions during those months.

Table 4.14 CVs for the regional unemployment data for 1986

	CV	*Lower*	*Upper*	*Months outside limits*
South-East	3.2	366	390	Jan.–Mar., Nov., Dec
East Anglia	3.5	81	87	Jan.–Mar., Oct.
London	1.5	402	414	Jan., Feb., Aug., Sep.
South-West	3.5	199	231	Jan., Feb., Jun.
W. Midlands	1.9	340	354	Jan., Feb., Sep., Nov., Dec.
E. Midlands	1.9	199	207	Jan., Feb., Jun., Oct.–Dec.
Yorks/Humbs	1.6	311	321	Jan., Feb., Sep., Nov., Dec.
North-West	1.9	440	457	Jan., Feb., Oct.–Dec.
North	2.6	229	241	Jan., Feb., Oct.–Dec.
Wales	3.4	173	185	Jan., Feb.
Scotland	1.6	354	366	Jan., Feb., May, Jun.

4.6 SUMMARISING RELATIONSHIPS

The graphs and summary measures presented so far are of value if one is interested in describing patterns in single variables. However, as the patterns found in one variable may correspond to those found in other variables, a natural development of the procedures outlined above is to search for patterns between variables. Such patterns may suggest relationships between the variables concerned which provide for a more effective understanding. Similarly, in searching for relationships, it may be possible to identify variables whose behaviour is particularly idiosyncratic, possibly indicating that their causal structures are independent of those found elsewhere.

4.6.1 Scattergrams

There are a number of different ways of describing and summarising relationships between variables using graphics and summary measures. The simplest graphical device available to describe relationships between continuous variables is the scattergram. This uses a two-dimensional graph to plot the observations of two variables with respect to each other. Figure 4.11(a)–(c) illustrates some of the types of bi-variate relationship which might be found in geographical data sets. In Figure 4.11(a) as the values of variable Y increase, so too do those of variable X. This indicates a positive relationship between the two variables. Figure 4.11(b) indicates a negative relationship because the values of X decrease as those of Y increase. In Figure 4.11(c), there is no easily discernible pattern. In such circumstances the two variables may be poorly related or not related at all.

Scattergrams are particularly valuable for displaying the ways in which the continuous variables vary with respect to each other. However, they may also be used to display scatter when one of the variables is categorical. Figure 4.11(d), for example, illustrates the differences in the level of

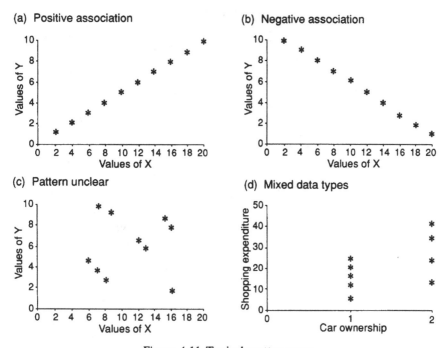

Figure 4.11 Typical scattergrams

shopping expenditure at an edge-of-town hypermarket between two groups of shoppers, one group which came to the centre by car (group 2), and the other which did not (group 1). The most obvious features of the information being portrayed here are the generally higher levels of expenditure among the car-borne group, and the wider scatter of shopping expenditure among the car-borne group.

4.6.2 Summary measures of relationships

A wide variety of summary measures have been devised to assess the presence and/or the strength of relationships between variables. Table 4.15 lists a selection of the more important of these, classified by the measurement scales of the two variables. As the two variables in Figure 4.11(a)–(c) are continuous, a suitable measure of their correlation is provided by the Pearson product moment correlation coefficient. Conversely, for Figure 4.11(d), where one variable is continuous and the other is categorical (nominal), their association might be assessed using the point biserial correlation coefficient or the difference of means test. Some of these procedures are used to assess hypotheses about the presence and/or strength of relationships between the variables. This approach will be considered in more detail in Chapter 6 and in Part II.

Table 4.15 Measures of correlation and association

Variable 1	Variable 2	Appropriate measures/tests
Binary	Binary	Difference of proportions
		Chi-square
		Fisher's exact test
		Yule's Q
		Kendall's tau
		Tetrachoric correlation*
Multi-way	Binary or	Chi-square
	Multi-way	Cramer's V
		Pearson's contingency
		Kendall's tau
Ordinal	Binary	Wilcoxon/Mann Whitney test
		Wald–Wolfowitz runs test
		Wilcoxon matched pairs test
Ordinal	Multi-way	ranked ANOVA
Ordinal	Ordinal	rank order correlation
		Kendall's tau
		Somer's D
		Wilson's E
Continuous	Binary	Difference of means test
		Point biserial correlation
Continuous	Multi-way	ANOVA
		Intra-class correlation
		Kruskal–Wallis 1-way ANOVA
		Friedman 2-way ANOVA
Continuous	Continuous	Pearson's product moment
		Linear regression

Note: * Useful if the underlying variables are categorised continua
Source: Adapted from Blalock (1979), Reynolds (1977), Cohen and Holliday (1982)

Before these measures are used to describe relationships it is important that some attempt is made to check that the data meet the assumptions associated with them. Assumptions are made to ensure that the hypotheses being tested can be assessed for significance against effects arising purely by chance. Without this check, it would not be possible to attach much reliability to the summary measures or the hypotheses they test. There are several different types of assumption, depending on which test is to be used. Most require that the variables are essentially random; others require that the variables come from a particular type of probability distribution. (Both aspects of these assumptions are taken up in Chapter 5.)

For an extended treatment of the use of these summary measures, see the classic series of papers by Goodman and Kruskal (1954, 1959, 1963, 1972), or the books by, among others, Siegal (1956), Reynolds (1977), Blalock (1979) and Cohen and Holliday (1982). The following subsections illustrate some of the characteristics of four of these measures.

4.6.3 Measuring correlation: two continuous variables

The Pearson product moment correlation coefficient (r) is a scaled index which measures the effect of paired variation between two continuous variables. It is written algebraically as:

$$r = \frac{\Sigma\,(x - \bar{x})\,(y - \bar{y})}{\sqrt{\Sigma\,(x - \bar{x})^2}\,\sqrt{\Sigma\,(y - \bar{y})^2}} \tag{4.1}$$

where the symbols refer to:

x	the individual observation on variable x
y	those for variable y
$\bar{x}$	the average value of x
$\bar{y}$	the average value of y
r	the correlation coefficient
$\sqrt{}$	the square root
Σ	summation symbol

The numerator (termed the covariance of the two variables) summarises the observed patterns of association between them. It reflects the fact that as values of one variable change those of the other may also change: both may rise together, fall together, or move in opposite directions. It is calculated as follows. First, re-express the individual observations of the two variables as deviations from their respective means. Second, multiply each pair of deviations together to form a product. Third, sum these products together to form a total. Fourth, divide this total by $N-1$. If most of the products calculated at step 2 are positive (negative), the overall covariance will be positive (negative). If, however, there is an equal number of both types the covariance will be roughly zero.

Covariance measures the bi-variate relationship in terms of the original units of measurement of the raw data. This means that it may be susceptible to change if the units are changed (see the discussion on interval and ratio measures in Chapter 3). In contrast, correlation measures the relationship in standardised units. This is achieved through the denominator in equation 4.1 which restricts the range of values obtained from the formula to the range −1 to 1. Values of −1 indicate a perfect negative relationship and 1 a perfect positive relationship. A value of 0 indicates that the two variables are not linearly associated. (This does not mean that they are not associated on any other scale.)

To illustrate the calculation of the correlation coefficient consider the data in Table 4.16 which come from the familiar six-times table. The figures labelled as variable Y are all six times larger than those in variable X (except when X and Y are both zero). The scattergram of this relationship is displayed in Figure 4.12. A number of features about this plot should be noted. First, if a line were to be drawn through all the

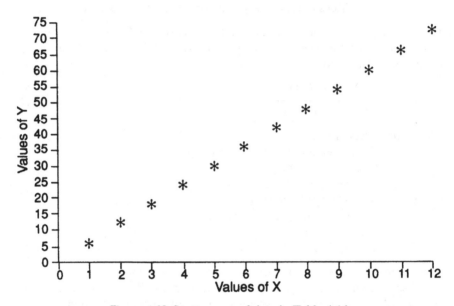

Figure 4.12 Scattergram of data in Table 4.16

crosses it would be completely straight. Second, this straight line would pass through the origin of the plot and the point representing the two means of X and Y (6.5 and 39). Third, the relationship between X and Y is constant along the whole length of the line, i.e., Y is six times larger than X both for small values of X ($X=1$) and for larger values ($X=9$). Fourth, the correlation coefficient for this relationship is 1.0.

The type of relationship being described by the correlation coefficient may now be set out in more detail. A correlation of 1.0 reflects that fact

Table 4.16 Six-times table

X	Y
0	0
1	6
2	12
3	18
4	24
5	30
6	36
7	42
8	48
9	54
10	60
11	66
12	72

that, when plotted, all the observations lie along a straight line and there is no scatter about the line. A correlation of 0.8 means that there is some scatter about the line but that this is small compared with the variation in the X and Y observations. The overall relationship of X and Y is thus nearly, but not quite, constant. A correlation of 0.3 indicates that the variation in the values of the X and Y observations is less than that in the scatter of points around the line. This means that the variation between X and Y is greater than that within either of them.

4.6.4 Measuring association: two nominal variables

Relationships between nominal variables may be summarised using the chi-square statistic. The main use of this measure is to describe relationships in contingency tables, i.e., tables created by the cross-classification of two or more categorical variables. Table 4.17 presents a simple illustration of this type of data. It contains the cross-classification of two categorical variables gathered as part of the Cardiff Consumer Panel Survey (Guy *et al*. 1983; Wrigley *et al*. 1985). Both refer to the use of superstores in Cardiff during the first six months in 1983. The column variable displays the pattern of patronage within the panel (coded as three sets – 'non-users', 'light' users, 'heavy' users). The row variable displays the pattern of car usage in grocery shopping within the panel (again coded as three sets – 'non'-users, 'infrequent' users, 'frequent' users). The table thus shows that 66 of the 451 panellists classified in the table as not using a car for shopping did not patronise superstores during the survey period. Similarly, 48 panellists who were heavy users of superstores frequently used a car for grocery shopping.

Table 4.17 Contingency table data

| | | SUPERSTORE PATRONAGE | | | |
		Nu	*Lu*	*Hu*	*Total*
	Non-users	66	71	29	166
CAR USE	*Infrequent users*	29	80	42	151
	Frequent users	25	61	48	134
	Total	120	212	119	451

Note: Nu – Non-users; Lu – Light users; Hu – Heavy users
Source: Adapted from Guy (1984)

The most obvious relationship to assess when dealing with contingency table data is the independence of the two classifying variables. For the data in Table 4.17, independence means that there is no difference in superstore patronage over the survey period between the three groups of car users. In other words, superstore patronage does not depend on consumers having access to cars. In order to assess this, a second contingency table needs to be generated in which the 451 shoppers are reallocated within the table to reveal the patterns which would be expected if the two variables actually were independent (Table 4.18).

Table 4.18 Expected values for the data in Table 4.17

| | | SUPERSTORE PATRONAGE | | | |
		Nu	Lu	Hu	Total
	Non-users	44.2	78.0	43.8	166
CAR USE	Infrequent users	40.2	71.0	39.8	151
	Frequent users	35.6	63.0	35.4	134
	Total	120.0	212.0	119.0	451

Note: Nu – Non-users; Lu – Light users; Hu – Heavy users

In the absence of other information the most likely number of frequent car users for the data in Table 4.17 is 134/451 or approximately 30 per cent. This means that for each class of superstore users, we would expect approximately 30 per cent to be frequent car users. The expected pattern of patronage would therefore be about 36 for non-users of superstores (i.e., 30 per cent of 120), 63 for light users (30 per cent of 212), and about 35 for heavy users (30 per cent of 119). Expected patterns for infrequent and non-users of cars may be generated in exactly the same way.

Having generated a contingency table of expected values, it is necessary to compare it with the observed patterns in Table 4.17. If the two classifying variables are indeed independent, then the observed and expected patterns should be similar. If they are not, then the two variables are either not independent or the data set is too small to make a fair assessment. This latter observation involves a statement of statistical inference: the subject matter of Chapter 6. The chi-square statistic allows this comparison to be made by calculating a measure based on the differences between the observed values and expected values at every point in the table. It is written algebraically as:

$$\chi^2 = \Sigma \; \frac{(o - e)^2}{e} \tag{4.2}$$

where

χ^2 is the calculated value of chi-square

o refers to the observed data values

e refers to their expected values under the hypothesis of independence

Σ is symbol for summation

Three steps are involved in its calculation. First, the difference between every observed and expected value in the table is calculated and the result squared. Second, these squared differences are divided by the expected values for each point in the table. Third, the nine resulting values are added to form an overall total.

If the two variables are independent the chi-square measure, generated as in equation 4.2, behaves in a way which is fully documented mathematically. As a result it is possible to distinguish with reasonable certainty between variables which are really independent and those which appear to be related but which, in reality, are not. For the data in Table 4.17, the chi-square measure is calculated as 28.57, which indicates that the two variables are not independent but are related, i.e., that superstore patronage does indeed depend on car usage. (The justification for this will be presented in Chapter 6.)

A variant of the inferential chi-square measure may also be used with the regional unemployment data of Table 4.1 to provide a consistent description of the performance of each region over the year. The motivation for this is that regional comparisons based only on the raw figures ignore the significance of unemployment within each region. As unemployment is likely to be larger in big labour markets such as the South-East and London, and smaller in areas such as East Anglia and the South-West, a comparison based solely on the raw data may be misleading. The 'signed' chi-square measure, so-called because the sign of the difference between the observed and expected values is reported, allows consistent comparisons to be made (Table 4.19).

The figures in Table 4.19 are calculated by comparing the observed figures for unemployment change in each region with an unemployment figure expected if the region mirrored the national change over the period. Negative figures are associated with regions which are performing better than the nation as a whole, and vice versa. Values outside the range plus

Table 4.19 Values of signed chi-square for regional unemployment change during 1986

Total	Change	% Change	Chi-square change
South-East	−33,058	−8.3	−320.7
East Anglia	−5,674	−6.5	−8.2
London	−15,033	−3.6	158.7
South-West	−14,766	−6.7	−32.2
W. Midlands	−19,852	−5.6	−0.1
E. Midlands	−11,101	−5.3	1.3
Yorks/Humbs	−14,530	−4.5	38.2
North-West	−27,009	−5.8	−4.1
North	−17,912	−7.3	−78.9
Wales	−16,822	−8.8	−219.7
Scotland	−5,900	−1.6	611.4

Note: Total change is the difference in unemployment as measured between January and December 1986

or minus 3.84 are considered to be significant differences. This means that the East and West Midlands both appear to be mirroring the national pattern of change, the North-West and East Anglia are performing slightly better than expected, and Wales and the South-East are performing very much better. Conversely, Yorkshire/Humberside, London and Scotland appear to be performing rather worse than that expected given the national picture.

4.6.5 Measuring rank correlation: ordinal variables

Relationships between two ordinal variables can be described using a form of correlation measure which recognises the ordinal or ranked structure of the variables. Two measures which may be appropriate are Spearman's rank order correlation:

$$r_s = 1 - \frac{6 \Sigma d^2}{n(n - 1)(n + 1)} \tag{4.3}$$

where,

    d is the difference in rank between the items in a pair
    n represents the number of items
    Σ represents the sum of the differences

and Kendall's tau:

$$\tau = \frac{s}{\frac{1}{2}n(n - 1)} \tag{4.4}$$

where,

    s corresponds to $P-Q$
    P is the sum of the number of ranks which are larger
    Q is the sum of the number of ranks which are smaller

Both measures require correction procedures to deal with ties – ranks on both variables which are equal (Cohen and Holliday 1982: 154–6 and 158–60).

The effect of ranking continuous data and then calculating a correlation coefficient can be illustrated using the data in Table 4.20, which is an extract of some precipitation data gathered for California (Taylor 1980). Columns 1 and 3 contain continuous measurements on precipitation (column 1) and altitude (column 3). The Pearson product correlation coefficient between these two variables is 0.592, indicating a moderately positive correlation between precipitation and increasing height. Columns 2 and 4 contain the ranks of the measurements in columns 1 and 3, with the largest readings being ranked 1 and the smallest 10. (The order of the rankings from top to bottom or vice versa does not affect the Spearman measure but is important for Kendall.) The Spearman rank correlation between these two variables is calculated as $1 - (612/990)$ or 0.382. The direction of the association is as given before but its strength has been reduced by the process of classification.

Table 4.20 Extract of Californian rainfall data

PPT	RP	ALT	RA
39.57	3	43	9
23.27	5	341	4
18.20	7	4,152	3
37.48	4	73	7
49.26	1	6,752	1
21.82	6	52	8
18.07	8	25	10
14.17	9	95	5
42.63	2	6,360	2
13.85	10	74	6

Note: RA and RP are the ranks of the ALT and PPT
    variables respectively
    PPT = Precipitation
    ALT = Altitude
Source: Taylor (1980)

4.6.6 Measuring association between mixtures of variables

The three previous examples described relationships between variables measured on the same scale. Social surveys invariably generate information at many scales and it is therefore important to know about measures which might help to describe relationships when the measurement scales are different. In Figure 4.11(d) the distribution of purchasing (a continuous variable) was related to the ownership of a car (a binary variable). To see if there is a relationship between these two variables, it is possible to use the point biserial correlation coefficient (r_{pb}):

$$r_{pb} = \frac{M_p - M_q}{SD} \sqrt{pq} \tag{4.5}$$

where,

M_p is the mean purchasing of car owners
M_q is the mean purchasing of others
SD is the standard deviation of total purchasing
p is the proportion of car owners to shoppers
q is the proportion of non-car owners to shoppers

Table 4.21 contains some hypothetical shopping data in which expenditure has been categorised by car-ownership. The five components of the measure are:

$$M_p = (24+14+36+41+25)/5 = 28$$
$$M_q = (13+17+21+24+7)/5 = 16.4$$
$$SD = 10.38$$
$$p = 0.5$$
$$q = 0.5$$

which produce a value of 0.558. This suggests that the association between expenditure and car-ownership is positive and moderately strong. One characteristic of this measure is that it may not reach either of its extreme limits (-1 and 1), especially if p and q are not equal. A test of significance is required to assess what it means. Once again, the justification for such a test is deferred until Chapter 6, but it can be shown that, at certain levels of significance, car-ownership does indeed exert a significant influence over shopping expenditures in this data set.

Table 4.21 Hypothetical shopping data

Customer	Spending	Car ownership
1	24	Y
2	14	Y
3	36	Y
4	41	Y
5	25	Y
6	13	N
7	17	N
8	21	N
9	24	N
10	7	N

Note: Y = Yes
      N = No

4.7 SOME FURTHER CONSIDERATIONS

The purpose of section 4.6 was to illustrate some of the summary measures which are available to search for and quantify associations or correlations between variables. Some of the measures presented are restricted in the number of variables they may handle, some are restricted to specific measurement scales, and some are restricted to both. Underlying measures such as the Pearson correlation coefficient and chi-square are assumptions that the variables being related are essentially random and independent. The importance of these will become clearer in Chapter 6, but for the moment it is sufficient to note that many geographical variables violate these assumptions. The effect of this is that the summary measures may be misleading and difficult to interpret correctly.

To illustrate the problem consider Table 4.22 which presents the Pearson product moment correlation matrix of the regional unemployment series in Table 4.1. The pattern of correlation between any two regions is symmetric so only half the table is printed. A brief inspection of this shows that the correlation values are positive and that:

1 Scotland correlates relatively lowly with every region, except perhaps the South-West.
2 London correlates relatively lowly with the exception of the South-West, Wales and East Anglia.
3 The patterns of correlations tend to reflect the spatial and structural positions of the regions concerned.

The latter implies that the unemployment values are probably not independent of each other spatially and, as we have already seen, there is considerable similarity in their values in successive months. When data of this sort exhibit structural relationships of any sort over space, time or both, the descriptive measures presented previously, and any models based on them, need to be modified to accommodate this.

Table 4.22 Correlation values for the regional unemployment data

	SE	EA	LD	SW	WM	EM	YH	NW	N	WL
SE	1.0									
EA	.96	1.0								
LD	.63	.41	1.0							
SW	.88	.89	.34	1.0						
WM	.82	.64	.89	.61	1.0					
EM	.97	.88	.76	.81	.91	1.0				
YH	.82	.71	.79	.64	.93	.91	1.0			
NW	.88	.77	.84	.66	.94	.95	.97	1.0		
N	.96	.93	.61	.81	.82	.95	.89	.93	1.0	
WL	.96	.95	.52	.91	.78	.93	.85	.86	.97	1.0
SC	.52	.45	.34	.76	.45	.54	.43	.42	.41	.53

The need to accommodate interrelationships between space and time (more commonly termed spatial autocorrelation) has been recognised by statisticians and geographers for many years. Initial efforts were aimed at devising measures for 'statistical maps' and essentially involved classifying the raw data into discrete categories and then counting the number of boundaries between zones in similar and dissimilar classes. Later work has been aimed at assessing autocorrelation through analyses of regression model residuals (see, for example, Cliff and Ord 1973, 1981) or correlograms (Cliff 1975). None of this work is particularly basic and much of it tends to be presented to students as a speciality for the more capable quantitative student. For a useful introduction to these issues, see Haggett *et al.* (1977: Chapter 11), Unwin (1981) or Upton and Fingleton (1989).

The works by Ripley (1981), Besag (1986) and Haining (1987) typify the sorts of research currently being produced by spatial statisticians.

4.8 SOME WORKED EXAMPLES

Most of the techniques presented in this chapter may be calculated using many standard computer packages. The MINITAB and GLIM 3.77 packages contain a reasonably comprehensive series of facilities for data description and analysis. As MINITAB is perhaps the most popular package used with introductory quantitative methods courses in geography, certainly in the UK, its use will be presented first. Data description facilities have been added to GLIM 3.77 as standard commands. Users with access to earlier versions of the package will not be able to use these.

4.8.1 MINITAB examples

The MINITAB package is particularly suitable for summarising and describing relatively simple data sets. It is command-driven and anticipates data entry in a case-by-variable data structure. This means that the rows in the data set correspond to geographical individuals (regions, rivers, people) and the columns to measurements made on them (unemployment, sinuosity, age). The data are entered into a worksheet and are then manipulated by applying commands to the columns. (It is possible to reverse this process and apply the commands to the rows, or to invert the rows and columns.)

The following command sequence might be used to analyse the East Anglian data in Table 4.1. (Lines 1 and 2 are operating system commands to set up a temporary external file to log the MINITAB session and activate the package.)

```
$empty -a
$run *minitab
outfile '-a'        set up a MINITAB log
read 'ea' c1        read data from file ea into column 1
print c1            print the contents of column 1
describe c1         generate summary statistics for column 1
histogram c1        produce a histogram for column 1
boxplot c1          produce a boxplot for column 1
stem-and-leaf c1    produce a stem-and-leaf plot for column 1
tsplot c1           produce a time-series plot for column 1
read 'sw' c2        read data for South-West into column 2
plot c1 c2          produce a scattergram from data in columns 1 and 2
correlation c1 c2   calculate the correlation coefficient
stop                terminate MINITAB run
```

The contents of file -a may now be inspected, copied to a permanent file for future reference, or routed to a printer.

The rounding associated with Ehrenberg's two-variable digit rule can easily be performed within MINITAB by dividing the contents of a variable by the required amount and rounding the observations to the nearest integer. The commands needed are:

divide c1 1000 c1   divide c1 by 1000 and restore in c1
round c1 c1   round c1 to nearest integer
describe c1   produce summary statistics on c1

4.8.2 GLIM 3.77 examples

The most recent release of GLIM, version 3.77, contains a limited amount of simple data description facilities to help users in a preliminary analysis of data. A useful introduction is provided in Swan (1986). Data may be entered from within the program or from external data files. All GLIM commands are prefixed by a system command which here is a dollar sign. Unlike MINITAB, the GLIM system allows greater flexibility over the form of the data which does not need to be in a case-by-variable format at entry. Indeed, the flexibility provided by GLIM arises because data structures can be created automatically within the package by sensitive use of the command language.

To process the data for East Anglia, it is first necessary to assign the twelve numbers for monthly unemployment to a GLIM variable. This can be done in a number of ways. One possibility is to create a variable EA and assign the values to it within GLIM: $ASSIGN EA=(data values)$. Alternatively, EA can be declared in a DATA command: $DATA EA$, and the numbers associated to it using $READ or $DINPUT. The former is used if the data are to be typed in at a keyboard, the latter if they already exist on a data file. In both cases, GLIM needs to be told how many items of data are to be input. This is done using the $UNITS command: $UNITS 12$. Once read in, the data may be displayed in a column by typing $LOOK EA$, or as a row by typing $PRINT EA$. Histograms are produced by typing $HISTOGRAM EA$, and manipulations, such as sorting into rank order, by $SORT REA EA$. In this case, the ranked values for the monthly unemployment have been sorted in ascending order and stored in variable REA.

Measures of central tendency and scatter can be produced using the $TABULATE command. This has a number of uses, so the specific function required is determined by a keyword. The following are some examples: $TAB EA MEAN$ – produces the mean value; $TAB EA FIFTY$ – produces the median; $TAB EA VAR$ – produces the variance, and $TAB EA DEVIATION$ – produces the standard deviation. Swan

(1986) shows how these may be used in conjunction with $CALCULATE commands to create cumulative frequency curves which are displayed using the $PLOT command.

4.9 SUMMARY

This chapter has outlined some of the steps required in providing a summary of a set of data. The key steps are:

1 Make a preliminary assessment of the data by data reduction or rounding to discern general patterns.
2 Calculate summary measures for central tendency and scatter.
3 Generate graphical displays for both central tendency and scatter to search for peculiarities in the data distribution.
4 Attempt to account for any irregularities which might be found.

These steps form a consistent core of procedures; they may need to be amended or extended depending on the nature of the analysis or the data type.

The reason for encouraging these steps in any data analysis is to minimise a poor interpretation of the data. There is now considerable evidence that a successful analysis cannot take place if only one or two of these steps are performed and the others excluded. This is because summary measures such as the mean and variance are affected by the shape of the data and may be drawn in value towards extreme data points. Conversely, the Pearson correlation coefficient can produce an identical summary value given radically different types of data distribution. The idea that a correlation value of 0.8 always means a straight line rising as the values of the variables get larger is not correct.

Whilst most of this chapter has been devoted to illustrating statistical procedures, it is important to realise that in many geographical data sets assumptions such as randomness may well be violated simply because of the structural nature of the information. The simple statistical procedures may on occasion be modified to account for some of the possible violations caused by geography, but the development of exclusively geographical measures is still a research issue. Until there is more general agreement about what constitutes geographical measures, and the software has been made available to implement them, most geography students will still rely on standard measures, modified wherever appropriate to geographical conditions. Underlying these measures are assertions about probability distributions and assumptions of randomness, independence and constant variance. These, and some of their variants, will be considered in subsequent chapters.

5

DESCRIPTIVE MODELS AND GEOGRAPHICAL DATA

5.1 INTRODUCTION

The measures and graphics presented in the previous chapter are particularly useful if the information in a data set is relatively simple to describe. Unfortunately, many geographical data sets are complex and require rather more sophisticated summary measures. This chapter describes some of the measures – so-called descriptive models – which may be helpful.

5.2 ALGEBRAIC EXPRESSIONS AS MODELS

Before presenting these models it is important to clarify what a 'model' means in this context, as it is a term which is frequently over-used. It is used here to refer to algebraic expressions which allow patterns or relationships in geographical phenomena to be exhibited, described and summarised. In this sense, 'models' should be viewed as extensions of the simple descriptive summary measures which were presented in the last chapter.

The central idea of a descriptive model may be illustrated by the data which make up the six-times table (columns X and Y in Table 5.1). From this, it is clear that Y is always six times larger than X for all values other than $X=0$. The scattergram of the relationship (Figure 5.1) shows that the crosses align perfectly. Similarly, their correlation coefficient of 1.0 indicates a perfect positive linear association. In summarising this information verbally, most researchers would be able to report without difficulty, that Y is six times greater than X for all values of X other than zero. Given this simplicity, it should not be difficult to see that the expression:

$$Y = 6X \tag{5.1}$$

conveys exactly the same information. This expression is thus a descriptive model of the relationship between X and Y in the observed data table.

A modification of this simple idea is provided by the relationships between X, Y, W and Z in Table 5.1. The scattergrams of the relationships

95

Table 5.1 Some linear relationships

X	Y	Z	W
0	0	15	15.0
1	6	21	19.75
2	12	27	24.5
3	18	33	29.25
4	24	39	34.0
5	30	45	38.75
6	36	51	43.5
7	42	57	48.25
8	48	63	53.0
9	54	69	57.75
10	60	75	62.5
11	66	81	67.25
12	72	87	72.0

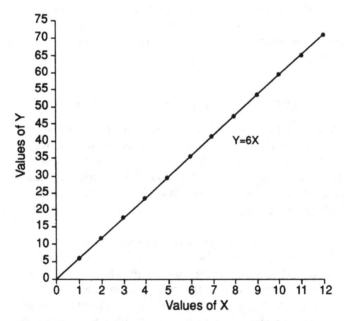

Figure 5.1 Scattergram of six-times table data

between X and the three other variables (Figure 5.2) show that though all three relationships are linear, they are subtly different. The relationship between X and Y is as given by equation 5.1; that between X and Z is similar, but the line cuts the vertical axis at 15 rather than 0. This produces a parallel line to XY which is 15 units farther up the Y axis. If the effect of this were removed, for example by subtracting 15 from every Z value, the new relationship would be identical to that between X and Y. As a

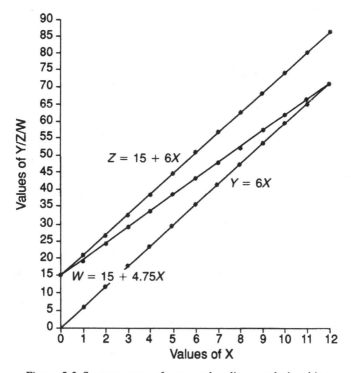

Figure 5.2 Scattergrams of some other linear relationships

result, equation 5.1 would apply. However, to indicate that when $X=0$, $Z=15$, a modification needs to be made to it:

$$Z = 15 + 6X \qquad (5.2)$$

where the 15 represents the fact that the XZ line lies 15 units higher up the vertical axis of the scattergram than XY. This new component of the model is frequently termed a constant, or an intercept, to reflect that it adds a constant value to Z for different values of X, and also, because it marks the point where the line 'intercepts', or crosses, the vertical axis.

The third line illustrated in Figure 5.2 (relating X and W) represents a relationship which seems similar to that between XZ for low values of X, but becomes more like that between XY for higher values of X. This line is parallel to neither of the others and so its equation cannot be deduced immediately merely from a quick visual inspection of the raw data. The equation for this line ($W = 15 + 4.75X$) has to be 'estimated' from the raw data. Some procedures of use for this task are introduced in Chapter 6.

Linear relationships, such as the three illustrated in Figure 5.2, can be described algebraically by a general equation for a line. This takes the form:

$$Y = \alpha + \beta X \tag{5.3}$$

where,

Y represents a 'dependent' variable, that is, a variable whose value depends on the model

X represents an 'independent' variable, that is, a variable whose value is fixed outside the model

α represents the value of the intercept

β represents the value of the slope coefficient

The exact relationship between X and Y depends on the values of the two unknown 'parameters', the intercept and the slope coefficient.

Equation 5.3 is of greatest use for describing relationships which are linear and additive and in which all the data points align exactly. However, this does not mean that the general model described above cannot also be used if some of the X and Y points deviate from a straight line, as illustrated in Figure 5.3. The effect of this situation is that no single linear equation will completely describe the data, as some deviation will always remain. However, it may be that one linear equation is considerably better than all the others and so could be used as a 'best fit' for the data. The level of inaccuracy arising from such a model may be so small as to make little difference to the eventual summary and interpretation of the data. (This important issue is considered in more detail in Chapter 6.)

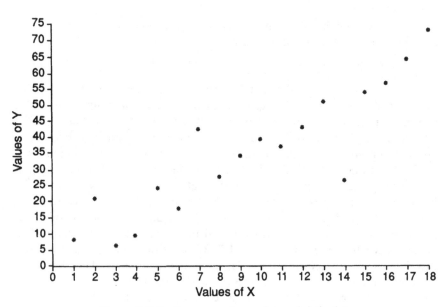

Figure 5.3 Scattergram of non-deterministic data

5.3 OBSERVED AND THEORETICAL FREQUENCY DISTRIBUTIONS

In a small and special data set such as Table 5.1, the descriptive model summarising the pattern between X and Y is relatively easy to see. However, in more complex examples some method of simplification may be needed. As with the histogram examples in Chapter 4, classification to exhaustive, mutually exclusive sets provides a way of simplifying the complexities of data without trivialising them. Here, such sets are termed 'observed frequencies', because they correspond to the frequencies with which members of the chosen sets are observed in the surveyed data.

Several different types of observed frequency distribution are likely to arise in histograms and stem-and-leaf plots of geographical data. The following are some common examples:

1 A (roughly) symmetrical distribution which peaks in the middle (Figure 5.4(a)).
2 A peaked distribution which may be skew towards one end (Figure 5.4(b)).
3 A J- or reverse-J-shaped distribution (Figure 5.4(c)).
4 A U-shaped distribution (Figure 5.4(d)).
5 A uniform distribution with equal frequencies for each observed value (Figure 5.4(e)).

It is possible to summarise the information in data sets distributed according to each of these shapes using standard summary measures. However, because of their sensitivity to extreme values, measures such as the mean and standard deviation are most effective when calculated on data which form symmetric curves (Figure 5.4(a)).

This does not imply that mean and standard deviation measures cannot be used with the other shapes. They may, so long as the frequencies being compared have the same shape and have similar amounts of scatter. This may be illustrated using the data in Figure 5.5, which show the number of times coffee was purchased by groups of 100 Cardiff households in four stores (a Leo's superstore, and supermarkets operated by Tesco, Kwiksave and Presto) over a 24-week period. The observed frequencies are roughly U-shaped, skew to the left, have mean levels of purchasing of approximately 3, and standard deviations of approximately 4. Thus, though the curves are not symmetric, it is still possible to indicate their similarities.

Knowledge that the observed values of a variable or data set form a particular shape can be helpful because it allows their possible description using 'theoretical' frequency distributions. Many observed frequency distributions can be simulated using data which have been generated artificially. The processes which are involved here are described in more

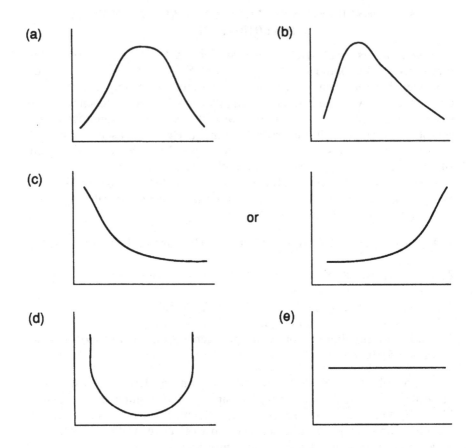

Figure 5.4 Observed frequency distributions
Note: Vertical axis: observed frequencies
Horizontal axis: histogram classes

detail in section 5.5. However, it is useful to note that many of the properties of data generated by these processes are known in great detail. As a result, if an observed frequency distribution can be approximated by an equivalent theoretical frequency distribution, its description may be considerably simplified.

Many different types of theoretical frequency distribution exist (Ord 1972; Johnson and Kotz 1969, 1970a,b, 1972), some of which are of value to the geographer. Those which are likely to be of most immediate value are: the Normal, the Poisson, the binomial and the multinomial. Each of these may be used to approximate observed frequency distributions, allowing more detailed descriptive information to be obtained from the observed data.

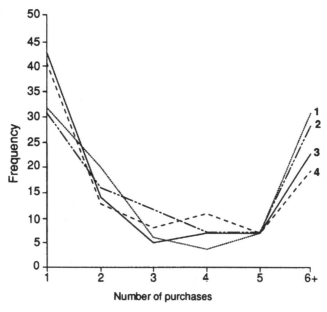

Figure 5.5 Observed purchasing of coffee
Note: 1–4 Each line represents individual stores

5.4 THE NORMAL DISTRIBUTION

Figure 5.6 illustrates the classic shape of the Normal theoretical frequency distribution. This distribution, which is also termed the Gaussian, is useful for describing the properties of continuously-distributed data. It has been particularly important in the history of statistics because it exhibits a number of attractive and powerful distributional properties. Moreover, it can be shown that under certain circumstances, many social and environmental phenomena tend to describe a Normal type of curve. However, symmetry and the characteristic bell-shape are only two of several properties which are exhibited by the Normal distribution. Other factors are also involved and, in practice, are rather more important. It is thus possible for symmetric, bell-shaped curves to be other than Normal.

5.4.1 Characteristics of a Normal distribution

A Normal distribution is described by an observed frequency distribution which possesses the following properties: symmetry about its mean; a peaked or 'bell' shape; identical values for measures of central tendency, and an identical pattern of scatter irrespective of the values of the mean or variance. This fourth property is particularly interesting because it

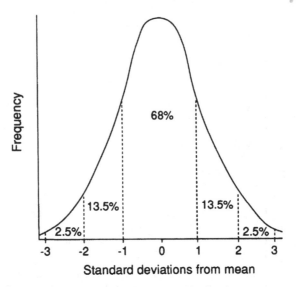

Figure 5.6 The Normal distribution

implies that the pattern of scatter under a Normal distribution will be the same irrespective of the summary measures. Specifically, 68 per cent (95 per cent and 99.7 per cent) of its observations will be distributed between the mean value and values located 1 (2 and 3) standard deviation unit(s) either side of it. These patterns are summarised graphically in Figure 5.6.

It is possible to show these patterns under different situations using data generated randomly from a variety of Normal distributions. Random numbers tables or computer packages such as MINITAB can be used for this task, though it should be noted that many random numbers generators are not totally random. This means that the patterns they display will be approximately Normal rather than fully Normal. To illustrate the idea consider the data generated in MINITAB using the following commands:

for variable 1: RANDOM 100 c1.
for variable 2: RANDOM 1000 c2;
              NORMAL 10 2.

The former generates a series of 100 random numbers from a Normal distribution with mean of 0 and standard deviation of 1, and stores them in variable c1. (This type of distribution is the default in MINITAB and will always be used unless overruled by sub-commands.) The latter generates a series of 1,000 random observations from a Normal distribution with mean 10 and standard deviation 2 and stores them in variable c2. (Notice the use of the semi-colon after c2, and the sub-command to override the default.)

102

The observed frequency distributions of both variables are displayed in Figure 5.7. As both are generated from (approximately) Normal distributions, they should exhibit approximately Normal patterns of scatter, irrespective of the fact that the number of observations involved, and the summary measures, are different. Table 5.2 compares the expected number of observations within 1, 2 and 3 standard deviation units of the two means with the numbers observed for the two variables. The similarity between the observed and expected figures for both variables is striking, even allowing for sampling variability. This leads to the general conclusion that the key factors in distinguishing between Normal distributions are the values of their respective means and standard deviations (or variances).

Table 5.2 Observed and expected numbers of observations

Within distance	Var 1 (n=100)		Var 2(n=1,000)	
	obs	exp	obs	exp
1 SD	70	68	680	680
2 SD	96	95	950	950
3 SD	100	100	997	997

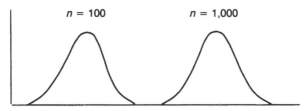

Figure 5.7 Different Normal distributions

5.4.2 Standard Normal variates

The random numbers generated by the default option in MINITAB (those with a mean of 0 and a standard deviation of 1 are frequently termed standard Normal variates or Z scores. They are of interest because they may be used to transform a Normal variate with given mean and variance into a second Normal variate with different mean and variance, and to assess simple relationships between observations arising from a Normal distribution. Z scores are calculated by subtracting the observed mean from each raw observation and dividing the result by the observed standard deviation.

5.4.3 Manipulating Normal variables

The ability to transform one Normal variable into the units of a second is frequently valuable if the research design requires that the observations

103

from different data sets be expressed in common units. This may arise in a variety of contexts and for a variety of reasons as, for example, in examination marking, where a bench-mark linked to age or ability range is required. The steps required to make this transformation are set out in Table 5.3 using the data for variable A. First, remove the effect of the original mean by subtracting it from each of the observations listed for variable A. This produces a series of 'differences' which have a mean of zero. These are listed in Table 5.3 under the heading 'Step 1'. Second, divide the figures listed under 'Step 1' by the standard deviation of the original observations for variable A. This removes the effect of their original scatter. The new values are z scores and are listed in Table 5.3 under the heading 'Step 2'. Third, multiply the z scores by the standard deviation required for the new variable, for example, 2. These are listed in Table 5.3 under the heading 'Step 3'. Fourth, add the mean value for the new variable, for example, 10, to the figures under 'Step 3'. These new figures are the raw observations for variable B, a Normal variable with mean 10 and standard deviation 2.

The ability to transform one Normal variable into a second suggests that the area under a Normal curve is organised systematically, linking together values of z and proportions of the data set. Thus, once we know that a variable is Normally distributed, we can automatically assume that 68 per

Table 5.3 Transforming one normal variable into another

	Var A	Step 1	Step 2	Step 3	Step 4
	13.4693	−0.41269	−0.28981	−0.57962	9.4204
	13.9130	0.03098	0.02176	0.04352	10.0435
	13.7557	−0.12625	−0.08866	−0.17732	9.8227
	13.7356	−0.14635	−0.10278	−0.20555	9.7944
	13.1781	−0.70389	−0.49431	−0.98861	9.0114
	14.3512	0.46921	0.32950	0.65901	10.6590
	13.7713	−0.11065	−0.07771	−0.15541	9.8446
	13.9860	0.10396	0.07301	0.14602	10.1460
	13.0440	−0.83796	−0.58846	−1.17691	8.8231
	13.1666	−0.71540	−0.50239	−1.00478	8.9952
	12.4173	−1.46474	−1.02861	−2.05721	7.9428
	11.1937	−2.68834	−1.88788	−3.77576	6.2242
	13.9080	0.02597	0.01823	0.03647	10.0365
	15.6039	1.72190	1.20920	2.41840	12.4184
	13.1030	−0.77904	−0.54708	−1.09416	8.9058
	16.4989	2.61689	1.83770	3.67540	13.6754
	12.1754	−1.70656	−1.19843	−2.39686	7.6031
	17.4954	3.61339	2.53749	5.07499	15.0750
	14.4350	0.55300	0.38834	0.77669	10.7767
	14.4410	0.55902	0.39257	0.78514	10.7851
Mean:	14.1560	0.0	0.0	0.0	10.0
SD:	1.5	1.5	0.0	2.0	2.0

cent of its observations will lie in a range 1 standard deviation unit either side of the mean. By extension, it is possible to calculate z for any given proportion and, conversely, any proportion for any given value of z. Table 5.4 lists the area under the Normal curve.

Table 5.4 Percentage areas under the Normal curve between mean value and given values of z

Z	0.0	0.1	0.2	0.3	0.4	0.5	0.6	0.7	0.8	0.9
0	0.0	4.0	7.9	11.8	15.5	19.2	22.6	25.8	28.8	31.6
1	34.1	36.4	38.5	40.3	41.9	43.3	44.5	45.5	46.4	47.1
2	47.7	48.2	48.6	48.9	49.2	49.4	49.5	49.7	49.7	49.8
3	49.9	49.9	49.9	49.9	49.9	49.9	49.9	49.9	49.9	49.9

Note: For $z = 1.3$, the area between the mean and z is 40.3%. To calculate the value either side of the mean, double the values in the table. For example, for $z = 1$ either side of the mean, the area under the Normal curve is 68.2% for different values of z and for different proportions.

The information in Table 5.4 applies to any Normal distribution and, consequently, is helpful in answering some simple questions about relationships between observed data. For example, if a Normal distribution has mean 100 and standard deviation 15, we know automatically that 68 per cent of the observations will lie between 85 and 115, 95 per cent will lie between 70 and 130, and 99.7 per cent will lie between 55 and 145. Values less than 55 and greater than 145 are generated very rarely from this Normal distribution, suggesting a way of checking for odd values in the data.

5.4.4 Non-linear transformations

Because of the simplicity of the Normal distribution many researchers tend to regard it as the key tool in describing data. Consequently, distributions which are non-Normal in shape are often transformed from their original units of measurement into alternative units, in the hope that these will be more Normal and consequently easier to describe. This approach is illustrated by the data in Figure 5.8 which are skew to the left, that is, most of the observations are smaller than the mean. Transformations which may be used with this sort of distribution are the square root and the natural logarithmic. These may be organised according to the strength of their influence on the original observations into a so-called 'ladder of powers' (Jones 1984). These, starting with the weakest, are: raw observations; square root; logarithmic; reciprocal root; reciprocal, and reciprocal square root. (Transformations for right skew are also available: square, cube etc.) The influences of some of these transformations on the skew data are summarised in Figure 5.9.

It is important to realise that not all skew distributions may be handled

Figure 5.8 Skew data

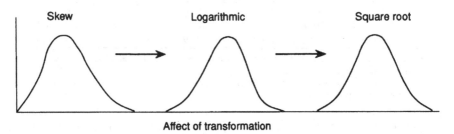

Affect of transformation

Figure 5.9 Effect of transformations on skew data

in this way. Frequently, the nature of the data, and the scale used in measurement, mean that a transformation is not appropriate. This is particularly so if the process of transformation produces a set of readings which are uninterpretable in terms of the physical or social processes being represented. Furthermore, it is rarely helpful if the data are categorical, for example, sex or mode of transport. The limited numerical structure of such data makes transformations impractical.

5.5 THE POISSON DISTRIBUTION

Many geographical variables are measured as nominal classifications or ordinal lists. As complex numerical operations are restricted with data measured on either of these scales (see the discussion in Chapter 2), it is often impractical to use the Normal curve to approximate such data. A more satisfactory approach is to use the Poisson, which Plackett (1974: 1) suggests 'stands in much the same relation to the analysis of categorical data as the Normal distribution does in relation to the analysis of measurements made on a continuous scale'. The Poisson is of particular use whenever the observations forming the data set have the following characteristics: they are counts; they are not restricted in their range of values (that is, they may range from zero to positive infinity); they occur independently of each other so that the occurrence of one observation does not affect the occurrence of any other; they occur irregularly without exhibiting a predetermined pattern; they occur at a constant

106

average frequency; and they are not restricted to a given number prior to analysis.

To illustrate its use, consider the following observed frequency distributions which show the classification patterns of two different types of count: the distribution of serious accidents over a thirty-week period (Table 5.5), and the distribution of flooding on a river over a period of fifty years (Table 5.6). Each table also lists a series of expected counts. These are the number of events expected in each measurement category assuming that the variables are from a Poisson distribution. Each of these has been generated from the following formula:

$$p_r = \frac{\mu^r e^{-\mu}}{r!} \tag{5.4}$$

where,

r represents the classes of accidents and floods
p_r represents the proportion of the data which takes the value ,
μ represents the mean value of the observations
r! is a term used to represent the product of r factorials (described in section 5.8.3)
e is a mathematical constant (2.718)

This is the characteristic formula for any Poisson distribution.

Table 5.5 Poisson distribution: distribution of serious accidents over a 30-week period

Accidents per week (r)	Observed	Theoretical
0	6	7.6
1	12	10.4
2	8	7.2
3	3	3.3
4	1	1.2

A simple inspection of the observed and expected values in Tables 5.5 and 5.6 shows that there is considerable agreement between them. This suggests that the Poisson distribution provides a reasonable description of the two series of observations. However, the Poisson distribution associated with Table 5.5 is not the same as that associated with Table 5.6. They differ because the values of the means of the two observed distributions are different (1.37 for Table 5.5 and 0.34 for Table 5.6).

Unlike the Normal distribution, which is fully described from knowledge of its mean and variance, the Poisson is described entirely from its mean.

Table 5.6 Poisson distribution: flooding
events over 50 years

Events per year (r)	Observed	Theoretical
0	35	37
1	10	10
2	4	2
3	1	1

This does not imply that variances cannot be calculated, they can. However, the Poisson possesses the special property that its mean is always numerically equal to its variance. The data in Tables 5.5 and 5.6 do not display this property exactly because of sampling variability. Consequently, both Poisson distributions will only provide approximate descriptions of their observed patterns.

Since the shape of a Poisson distribution depends entirely on the value of its mean, it is possible to generate a variety of Poisson distributions each based on a different mean value. However, a number of general results for Poisson distributions should be noted (these are based on data sets containing fifty observations):

1 If the mean value is less than 1, the Poisson will describe a reverse-J shape similar to Figure 5.4(c).
2 If the mean is between 1 and 4, the Poisson will be hump-backed in shape, but skew to the left (Figure 5.4(b)).
3 If the mean is greater than 4, the Poisson will be roughly symmetrical and hump-backed (Figure 5.4(a)). Poisson distributions which are shaped in this way are very similar to Normal distributions. It is thus possible to describe their structure as a special case of the Normal distribution in which the summary variance measure is equal to the mean.

These results are summarised graphically in Figure 5.10.

The traditional role of the Poisson is to describe patterns among so-called rare events, for example, natural hazards, strikes, or accidents. In geography, considerable attention has been paid to using the Poisson to describe patterns among point data (see Chapter 2). A typical example is in settlement studies and locational analysis where a quadrat or grid is placed over a map, and the number of settlements (or some other phenomenon) is counted. Modifications of the basic Poisson model allow different types of settlement process to be investigated. For further details, see King (1969), Cliff and Ord (1973), or Haggett et al. (1977).

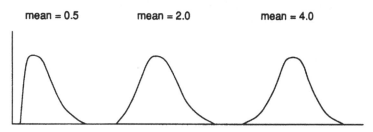

Figure 5.10 Various Poisson distributions

5.6 THE BINOMIAL AND MULTINOMIAL DISTRIBUTIONS

The Poisson distribution provides a powerful descriptive tool for analysing counted data. However, two conditions may make its use unreasonable. First, the researchers may decide that the selection of observations has to be limited to a fixed number, perhaps to meet cost or labour constraints. Second, the research design chosen may require the simultaneous measurement of the frequency distributions of more than one type of event.

The first condition is frequently met as resources are nearly always limited in research and operational decisions have to be made about how much data may be collected and analysed. By fixing the size of the measurement sequence to a specific length, the researchers violate one of the key assumptions of the Poisson. The second problem generalises the research interests, but again violates the Poisson, which is concerned with the frequency distribution of one variable only. In order to accommodate these new requirements, some alternative discrete distributions should be used instead. The two which are of most immediate value for these problems are the binomial, which is appropriate when two distinct types of observation are to be considered, and the multinomial, for problems involving more than two types of observation. (For details of the multinomial, see Dobson 1983.)

In order to make use of both distributions, the data set must exhibit the following features. First, it must be made up of a number of sets or subgroups of observations which are of fixed length. Second, the observations in each group must occur independently of each other, be irregular in value, and occur at a constant average rate. Third, they must be classified into two (binomial) or more (multinomial) exhaustive and mutually exclusive categories (for example, measurements recorded as 1 represent the presence of the event, whereas those recorded as 0 represent its absence).

The distribution of the observed proportions of events is displayed in Figure 5.11. To describe this using the binomial or multinomial

109

theoretical distributions, an equivalent series of expected proportions needs to be generated. For a two-event problem, this may be done using the characteristic equation for the binomial:

$$p_r = \frac{n!}{(n-r)! \; r!} \; p^r \, (1-p)^{n-r} \qquad (5.5)$$

where,

p   refers to the average proportion of events of type r in the data set

$1-p$ refers to the average proportion of events of type $n-r$ in the data set

p_r refers to the proportion of groups in the data set which would be expected to contain r events if they were binomially distributed

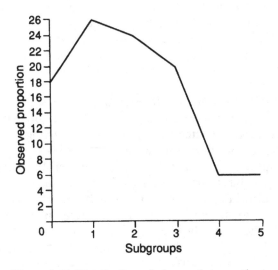

Figure 5.11 Distribution of observed proportions

To illustrate the basic idea, consider the data in Table 5.7 which reflect the observed results of an hypothetical survey conducted to gauge public opinion on the future building of nuclear power stations in Britain. Assume that the data set was collected from across the country by a series of interviewers, each collecting twenty interviews. The table shows the observed proportion of responses in favour of future building across all the fixed length groups of interviews. The observed data show that 18 per cent of the groups contain no favourable responses, 26 per cent contain one favourable response, 24 per cent contain two, and so on. The average proportion of favourable responses from all groups (p) is approximately 0.1 (corresponding to 97 positive replies out of 1,000 interviews).

Table 5.7 Binomial distribution: proportion in
favour of nuclear power stations

Proportion	Observed %	Theoretical %
0	18	12
1	26	28
2	24	30
3	20	20
4	6	6
5	6	4

The equivalent proportions which would be expected were the observed proportions to follow a binomial distribution are also listed in Table 5.7. These are calculated as follows: for groups of twenty interviews containing five positive responses the expected binomial proportion is calculated by substituting 20 for n, 5 for r, 0.1 for p and 0.9 for $1-p$ in the binomial equation. Thus, in a binomial distribution, 4 per cent of the groups containing twenty observations would have five favourable responses and fifteen unfavourable responses. This seems to be very close to the observed proportions. The fit of the binomial to the other groups also appears to be reasonable, which suggests that a suitable way of describing the observed data is to say that they are binomial in form with an overall proportion in favour of nuclear power of one in ten ($p=0.1$).

Unlike the Poisson, the mean and variance of a binomial variable are not assumed to be equal. Both may be calculated directly from the data using modifications of the formulae for grouped data which were presented in Chapter 4. However, a theoretical short cut is also available. Using this, the mean may be calculated directly from the expression:

mean = np

and the variance and standard deviation from the expression:

variance = npq
standard deviation = $\sqrt{(npq)}$

(where $q = 1-p$). Thus for the data in Table 5.7, the mean is calculated as $20 \times 0.1 = 2$, the variance is $20 \times 0.1 \times 0.9 = 1.8$, and the standard deviation is 1.34.

The binomial distribution is fully described once the size of the fixed length groups and the overall proportion in favour are known. Its shape, however, depends on their actual values. Figure 5.12 presents a number of different binomial distributions based on fifty randomly-generated samples whose shapes reflect the relative sizes of n and p.

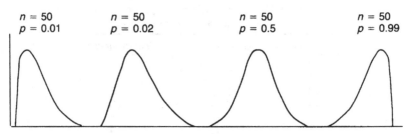

Figure 5.12 Various binomial distributions

5.7 RELAXING SOME ASSUMPTIONS AND GENERALISING

The theoretical frequency distributions presented in the previous sections may be considered a 'basic set', that is, a small group of powerful, general distributions. Whilst they are of considerable value in approximating many types of data, it is quite likely that there will be situations where their assumptions will be too restrictive. A variety of modified theoretical distributions can be produced by relaxing some of these assumptions. These include the negative binomial which, in spite of its name, is a modification of the Poisson, and the beta-binomial, which is a modification of the binomial.

In addition to these modified distributions, a variety of generalised distributions also exists which can be shown to be related to members of the basic set. The value of these will become clearer in the context of hypothesis testing, which is introduced in Chapter 6, and in Part II. These include the chi-square distribution, the *t*-distribution, and the *F* distribution. This section presents some outline information on these and the two modified distributions listed above.

5.7.1 The negative binomial distribution

The negative binomial distribution is generated by modifying one of the assumptions of the Poisson distribution. In a Poisson distribution, the mean value at which events occur is assumed to be a constant. Each element in the data set is thus assumed to have been generated at this constant mean rate. However, in the negative binomial, this assumption is relaxed by assuming that parts of the data set may have been generated at different mean rates of occurrence.

One important use of this sort of distribution occurs in market research where retailers are interested in describing the purchasing patterns of households over time. It can be shown that the purchasing behaviour of an individual household through time will follow a Poisson distribution (Ehrenberg 1972; Sichel 1982), but that different households will purchase

112

products at different mean rates. To describe a group of households using the same Poisson distribution will thus result in a poor fit between observed and expected values and consequently a poor description of the data. A negative binomial distribution offers the possibility of a much better description and fit simply because it can make allowance for the different mean rates of purchasing.

A good example of the use of the negative binomial in a geographical context is provided by Dunn *et al.* (1983) and Wrigley and Dunn (1984a) who used it to describe the patterns of shop visits in the Cardiff Consumer Survey. They found that the negative binomial applied to household purchasing and travel data provided reasonably acceptable descriptions of repetitive brand purchasing and store patronage. And in addition, the descriptive model was able to differentiate between purchasing behaviour at central stores with city-wide market areas, and stores located in peripheral sub-areas of the city.

The negative binomial has also been widely used by geographers and other locational analysts to study assumed contagious processes, for example, the spread of a disease, or the growth of a settlement pattern over virgin territory. Haggett *et al.* (1977) discuss at length some of the problems encountered in applying this sort of distribution to grid-square data gathered in Japan (see also Cliff and Ord 1973). They show that a family of Poisson-based models, including the negative binomial, may be used to approximate the settlement changes and proximity relationships represented in the grid squares, and can be developed to accommodate different mean rates of change within the spatial system.

5.7.2 The beta-binomial distribution

The beta-binomial distribution is generated by modifying one of the assumptions of the binomial distribution. Once again the assumption which is relaxed concerns the average rate of occurrence of the event being studied. The binomial assumes that the events occur independently, irregularly and with constant mean rate within groups of observations of fixed length. The beta-binomial relaxes this to allow variable mean rates of occurrence to be described. A geographical example of its use is given in Dunn and Wrigley (1985), in which an attempt is made to describe the patterns of patronage of a series of competing shopping centres in Cardiff. The extension of this distribution to handle variable rates of patronage in heterogeneous data sets (that is, data sets in which consumers with identical attributes such as age or mobility behave differently) and multi-way choices is currently a major research topic. Some of the issues are reviewed in Uncles (1988).

5.7.3 The chi-square distribution

The z score, or standard Normal variate, was presented in section 5.4 to illustrate some of the attractive properties of the Normal distribution. It may also be used to describe a second distribution based on a series of independently distributed standard Normal variables. This distribution is termed the chi-square distribution; a skew, continuous distribution whose values range from zero to positive infinity, and whose shape depends on the number of independent standard Normal variates 'degrees of freedom' involved (Figure 5.13). The value of this distribution for assessing ideas about relationships in data sets will become clearer in Part II, where it provides the bench-mark against which models for categorical and continuous data may be compared.

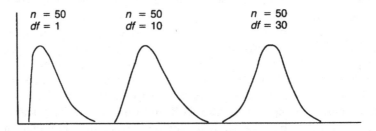

Figure 5.13 Various chi-square distributions

5.7.4 The *t*-distribution

The *t*-distribution is a continuous distribution which, like chi-square, is primarily used in the assessment of hypotheses. It is formed by the mixture of Normal and chi-square distributions, and is most valuable when the number of observations in the data set is considered to be small. The shape of the *t*-distribution becomes increasingly more Normal as the number of degrees of freedom increases. This relationship has meant that *t* is most frequently used to assess hypotheses about Normal variables whose variance is not known (see Chapter 6).

5.7.5 The *F* distribution

Like chi-square and *t*, the *F* distribution is a continuous distribution most frequently associated with assessing hypotheses. It is defined as the ratio of two chi-square variables, each divided by the number of independent terms associated with them. It is most widely used in studies where the variability in a data set can be decomposed into distinct subsets attributable to clearly-defined sources. The purpose of this is to assess the relative influence of each of these sources on the overall information in the data set. This distribution will be considered in more detail in Part II, in particular in connection with the analysis of variance and regression.

5.7.6 Interrelated distributions

The fact that so many of the theoretical distributions appear to describe similar shapes under suitable conditions suggests that they may substitute for each other. The t and chi-square distributions provide useful alternatives to the Normal and can be suitable ways of assessing if a variable is indeed Normally distributed. However, it can also be shown that the Poisson, binomial and Normal distributions mirror each other. If the value of p in a binomial distribution is so small that $1-p$ is approximately 1, then the binomial mean and variance will be almost identical. As this is a property of the Poisson, it follows that the Poisson may be used to represent binomially-distributed data of this form. Similarly, if the binomial is symmetric in shape, it can prove acceptable to use the characteristics of the Normal distribution to describe binomial data. The approximation of the two distributions improves with the size of the data set.

This leads to a major problem. As competing descriptions of the raw data using different distributions are possible, it is frequently difficult to determine which description is correct. To tackle this problem it is necessary to investigate the potential processes underlying the data.

5.8 PROBABILITY DISTRIBUTIONS

5.8.1 Probability and theoretical frequency distributions

The previous three sections have introduced the idea that it is possible to describe observed frequency distributions using a variety of theoretical frequency distributions. These were shown to vary in shape and form, and to be appropriate for a variety of types of data and ranges of numbers. As these theoretical distributions are interrelated, it is reasonable to assume that they may share some common basis, that is, that they may all arise in a similar, if not identical, way.

In order to understand the basis of these theoretical frequency distributions it is necessary to consider the idea of probability and how it may refer to a single observation in a data set. Probabilities are used to provide measures of the chances of occurrence of particular random, or approximately random, events. The term 'approximately random' is used to reflect the fact that 'true randomness is only an abstract concept' (Ehrenberg 1982: 75), and that it is not possible to be absolutely certain that a specific observation is random. This aside, the main use of probabilities is to calculate the chances of a particular type of event occurring if it were essentially random. This measure will depend on the nature of the data, the type of study being conducted, and the number of observations in the data set. It will also depend on the particular probability process which is being considered.

Each of the theoretical frequency distributions presented previously depends on the existence of an underlying probability process. The Poisson is generated by a Poisson probability process; the binomial, by a binomial probability process, and so on. Each theoretical distribution summarises the shape that a set of data would exhibit if it were generated by a particular process. It is thus possible to use the idea of a probability process to provide a possible explanation for the form of any observed data which may be described by a theoretical frequency distribution.

To illustrate the relationship between a theoretical distribution and a probability process consider the case of the Normal distribution. This exhibits some particularly simple descriptive properties, for example, that 95 per cent of all observed items lie within 2 standard deviations of the mean. If the theoretical frequency distribution represents the shape of any series of n independent Normal events, then it can be argued that the probability of any individual observation being within 2 standard deviations of its mean value, whatever that happens to be, is $p = 0.95$. In effect, the theoretical distribution summarises the effect of the process in aggregate; the probability process summarises the relative probabilities of any event generated by that process actually occurring.

5.8.2 The Normal process: Central Limit Theorem

The characteristics of a Normal distribution are automatically produced whenever observations in a data set are independent of each of other, produce additive effects, are irregular in numerical value and can range from minus infinity to plus infinity. It can be shown that in aggregate, these conditions always lead to the classical Normal shape. The process which ensures that this occurs whenever the generating conditions are correct is termed the Central Limit Theorem. This provides a potential mechanism for explaining the structure of many different types of process, including those which are themselves non-Normal (see section 5.7.6). Probabilities are generated from a Normal process according to the following characteristic formula:

$$e^{-(x-\mu)^2/2\sigma^2} \tag{5.6}$$

where,

    e represents the base of natural logarithms (2.718)
    x represents each observation
    μ represents the mean value of the data
    σ^2 represents the variance

The two terms μ and σ^2 characterise each Normal distribution, a feature which was emphasised in section 5.4. Such terms will be referred to in Part II as canonical parameters.

5.8.3 Some other processes

Other probability processes are produced if the conditions underlying the Central Limit Theorem are not met. For example, the conditions required for the generation of a Poisson process are as follows. First, the observations must be counts which are independent and random. Whereas the Normal process expects data to range from negative to positive infinity, Poisson data are generally restricted to values of zero or greater. Second, the size of the data set must not be fixed in advance. Third, the observations must occur with constant probability. Similarly, the binomial process applies if the observations can only take on two distinct values (A and B). It is generated if each outcome can be shown to occur irregularly, and is an independent event which is based on a stable probability of occurrence.

5.9 SUMMARY

The ability to describe patterns and relationships between geographical phenomena algebraically offers researchers a considerably more powerful descriptive tool than the simple summary measures of Chapter 4. Two types of descriptive model were presented here. First, models based on linear additive relationships. Second, models based on theoretical frequency distributions. Both of these provide ways of summarising patterns in terms of simplified algebra, allowing similarities to be emphasised, and oddities to be illustrated. Apart from the trivial cases where the relationship is exactly linear, or where the data are generated by a specific probability process, all models approximate the observed data, that is, they account for as much of the variability as is possible without necessarily accounting for it all.

The problem with this is that the variability omitted may actually be important, leading to potentially serious misinterpretation if ignored. As a result, some mechanism needs to be developed which can accommodate model inaccuracy explicitly, that is, without relegating the departures from the model to the status of by-products. The handling of this problem varies greatly, depending on which type of model is being used. However, because many probability processes and theoretical frequency distributions are interrelated, a consistent approach may be developed which accommodates most of the problems likely to be encountered, at least in common analyses.

This argument is taken up in more detail in Part II. However, before

117

proceeding to this, it is necessary to review one final component of data analysis: generalising the results from the confines of the specific study to the wider world. Two approaches offer help here. One develops the data reduction procedures suggested by Ehrenberg and introduced in Chapter 4. The other makes use of statistical inference, and the somewhat amazing finding that there is frequently sufficient information in a single observed data set both to describe its relationships and indicate its relevance to the wider world.

5.10 SOME FURTHER NOTES ON PROBABILITY

The value of probability-based descriptive models is that they provide a particularly useful link between the simple summary measures considered in Chapter 4 and the inferential statistical procedures which follow in Chapters 6–10. There is a clear parallel between probability and statistical inference. In any situation where one is concerned with investigating the structure of collected data, attention can be focused on either the possibility that the structure in the data set arises by chance (a probability question), or assessing the chances that a specific type of data could arise when their probabilities are not known (a statistical inference question). Chatfield (1983: 37–8) attempts to summarise this as follows:

> The duality between probability and statistics can be expressed in the following way. In both subjects the first step in solving the problem is to set up a mathematical model for the physical situation which may involve one or more [terms] ... If the [terms] of the model are known, given or established from past history then we have a probability problem and can deduce the ... [structure of the data] ... from the model. However if the [terms] are unknown and have to be estimated from the available data then we have a statistical problem.

Inferential statistics, especially classical theory, is based on the properties of probabilities. Many of the most important results which were used to develop the generalised family of linear models were based on probability theory.

The classic description of probability theory is based on the so-called frequentist argument (see, for example, Silvey 1975). This postulates that if an event is true its probability will stabilise in value if sufficient information is collected. For example, if sufficient throws are made of an unbiased coin, the proportion of heads and tails observed will be roughly equal. The logic of frequentism is based on three general rules:

1 the number of possible outcomes (heads or tails) is clearly defined and limited in number;

2 the coin is not 'loaded', for example by not having two heads or tails, or an inbuilt tendency to favour one outcome;

3 each throw of the coin is independent of any other throw. This means that the outcome of one throw does not affect the outcome of any other throw.

If these hold, then in the long run, the percentage of heads and tails will be approximately 50 per cent and the respective probabilities approximately 0.5.

The coin-tossing example characterises the principal ideas in a simple probability problem. First, there is a variety of possible outcomes, but which will occur on a given occasion is not known. Second, the range of possible outcomes is known with freak events such as the coin landing on its edge being discounted. Third, the mechanism used with each toss of the coin is assumed to be the same. In the language of statistics, a single toss is called an 'experiment', the two possible outcomes, 'sample points', and the set of possible outcomes, the 'sample space'. In order to apply this form of logic to an alternative problem therefore, researchers need to clarify all three of these features.

5.10.1 Permutations and combinations

The mathematics of permutations and combinations may be used to assess the structure of the sample space and the character of each of the sample points. The number of ways in which specific types of event, r, can be selected from a given number of items, n, can be expressed algebraically as:

$$\binom{n}{r} = \frac{n!}{r!\,(n-r)!} = \binom{n}{n-r} \tag{5.7}$$

where

r represents the number of events of given type
n represents the overall number of events
! is a symbol representing a factorial structure ($n! = n\,(n-1)\,(n-2)\ldots$)

To illustrate this let $n = 5$, $r = 3$ and the object is to see how many combinations of $r = 3$ can occur in sequences of $n = 5$ items:

$$\binom{5}{3} = \frac{5!}{3!\,2!} = 10 \tag{5.8}$$

In this equation no attempt is made to assess the order of the selection of the $r = 3$ items. If the r items of n were the letters a–e, then the ten combinations of any three letters would be: *abc, abd, abe, acd, ace, ade, bcd, bce, bde* and *cde*. If order is important, we need to calculate the

119

number of permutations of size r available in any sequence of size n. The algebra for permutations is:

$$np_r = \frac{n!}{(n - r)!}$$ (5.9)

which, for $n = 5$, $r = 3$ is:

$$5p_3 = \frac{5!}{2!} = 60$$ (5.10)

yielding the following:

abc bac cab dab eab abd bad cad dac eac abe
bae cae dae ead acb bca cba dba eba acd bcd
cdb dbc ebc ace bce cbe dbe ebd adb bda cda
dca eca adc bdc cdb dcb ecb ade bde cde dce
ecd aeb bea cea dea eda aec bec ceb deb edb
aed bed ced dec edc.

Combinations such as *abc*, *bca*, and *cab* are identical, consisting of different arrangements of the same letters. However, they are different permutations. Each permutation or combination represents a single sample point; collectively all permutations or combinations represent the sample space. As a result, the probabilities associated with obtaining any single combination or permutation of size $r = 3$ when $n = 5$ are 0.1 (combinations) and 0.016 (permutations). In other words, these probabilities measure the relative chances of any particular event, or type of event, occurring compared with the total number which could occur.

5.10.2 Types of event

Two types of event may be considered: mutually exclusive events, and non-mutually exclusive events. In the former, if A and B are two possible outcomes, mutual exclusivity implies that if A occurs then B cannot, and vice versa. The latter case covers a wider range of types of event. For example, A and B occurring together, A and B being independent of each other, and A and B being dependent on each other in some way.

Models of independence and dependence represent two types of conditional probability, that is, probability statements which presume that certain things happen. Thus if one is interested in knowing the numbers of combinations between any two letters once the letter A has been selected, the problem focuses on only those events in the sample space containing A. In other words, the probability is conditional on A having been selected. This amounts to restricting the sample space from ten events to six in the example given previously. If two events are independent, it follows that the fact that A is selected does not influence the selection of B.

5.10.3 Other bases for inference

The frequentist approach to probability is particularly important in developing a basis for inferential statistics but it is not the only such basis available. An alternative probability approach, termed Bayesian analysis, makes use of so-called subjective probabilities. In this, prior information or existing knowledge is used to estimate a series of so-called prior probabilities; probabilities which represent what the researchers think might happen. After the experiment, these priors are modified by experience into posterior probabilities; probabilities based on what actually happened. The purpose of the approach is to capitalise on accumulated knowledge. However, though mathematically unflawed, there is considerable argument about how to calibrate the prior terms in the first place. The dispute is currently unresolved.

An entirely different inferential system also exists. This is termed likelihood inference and is discussed in detail in, among others, Edwards (1972), Sprott (1973), Silvey (1975) and Pickles (1986). The purpose of this procedure is to base inference firmly on the information contained in the observed data without appealing to any hypothetical long run. As the possible sources which could have generated these data are limited, it is argued that by maximising, in some way, the information in the data, the characteristics of the source become more evident. This type of inference is used extensively in Part II, and is introduced in Chapter 6.

6

STATISTICAL INFERENCE

6.1 INTRODUCTION

The aim of the previous chapters has been to describe and summarise patterns in single sets of data. Until now, we have assumed that these patterns are specific to those data sets. However, it is possible that certain aspects may be more general and might emerge again if other data sets were collected from the same geographical individual. This tendency for data to possess both common and idiosyncratic patterns is important in statistics because it provides a means of studying large and complex 'populations' without enumerating all of their characteristics. This process is termed 'statistical inference', and is the focus of this chapter.

6.2 STATISTICAL TERMINOLOGY

The terms 'population', 'parameter', 'sample', 'statistic', and 'inference' have specific meanings when applied statistically. The following are some typical definitions of each term:

1 *Population*:   a well-defined collection of observations, objects or events which have one or more characteristic properties in common.
2 *Parameter*:   a numerical measure which describes some of the properties of the population.
3 *Sample*:   a representative subset of a population which is often used in analysis as a proxy or substitute for it.
4 *Statistic*:   a numerical measure which describes some of the properties of a sample.
5 *Inference*:   the process of estimating the properties of population parameters from sample statistics to within predictable levels of accuracy.

The following subsections expand on these definitions in greater detail.

6.2.1   Population and parameters

In data analysis, the term 'population' is used to refer to any collection of objects, observations or events which share some common, measurable characteristic(s), and which is to be used as the focus of study. For example, in an investigation of the UK higher education system the total number of students enrolled on courses could be used to define the population for the study. Similarly, market researchers may define a population using location and the socio-economic status of consumers in order to target an advertising campaign designed to sell specialised services. This population would exclude consumers living outside the chosen areas as well as those within the areas who were in different socio-economic groupings from those required. The identification of a population is therefore a classification problem similar in nature to those introduced earlier in Chapter 3.

As a result, each member of a population exhibits the same defining characteristic(s); that is, all members are students or are residents in the target area. However, this does not mean that these members are identical. In the educational population presented above the defining characteristic for membership is enrolment on a course of higher education. Though all members possess this characteristic, they may differ in many key ways. For example, they may differ in age and sex, be taking different types of course (first degrees, higher degrees, diplomas etc.) in a variety of different subjects, and be following different types of study (full time, part-time, sandwich courses). These differences may be measured and recorded numerically as 'variables', i.e., a selection of measurements which illustrate the range or the variability of a specific characteristic (e.g., age) within the population.

Many populations are numerically large. For example, the national population of Great Britain as measured by the 1981 Census is about 54 million people. Similarly, the population of persons unemployed during 1986 defined by official sources is about 3 million. As a result of this, it is impractical to attempt to describe the variables present in such populations person by person. A more useful approach is to calculate a set of summary measures such as the mean and variance (or any of the other measures described in Chapter 4) and use these instead as proxies for the population. When calculated on population variables, such summary measures are termed 'parameters'.

6.2.2 Samples and statistics

The magnitude of a population often raises a second difficulty: it is frequently too large to be studied in its entirety. This is true even for the calculation of parameters, most of which tend to be based on the manipulation of considerable quantities of data. Accordingly, unless the project requires information relevant to each specific member of the population, a more tractable strategy is to select a representative subset of it which may be analysed instead. Such a subset is termed a 'sample'.

Samples are widely used in social and environmental research to provide the data needed to study social or environmental populations. For example, the voting behaviour of the British electorate is frequently gauged using opinion polls of the voters in a small number of constituencies located around the country. These may be chosen to include urban and rural constituencies, northern and southern locations, Conservative and Labour areas, and so on. Similarly, the short-term variability in grocery prices is frequently gauged by pricing a selection of groceries in a hypothetical, 'typical' shopping basket on the same day each week in a number of different stores within a city or around the country (Guy and O'Brien 1983; O'Brien and Guy 1985). The same idea also applies to research problems in physical geography. For example, Anderson (1977) and Anderson and Cox (1986) show that the pattern of movement on a hillside may be gauged over time by comparing changes in the relative positions of a small number of pegs which have been hammered into it to act as indicators of movement. Collectively, the pegs act as a representative sample of sites on the hillside which could have been chosen for the study. By focusing on them, information on movement across the whole hillside may be obtained without the need for more detailed monitoring.

As with population variables, the defining characteristics associated with sample members will not be identical but will vary. This means that some attempt must be made to describe their variability. This may be done in exactly the same way as for populations by the calculation of summary measures such as the sample mean and variance. To distinguish these summary measures from population parameters, sample summary measures are termed 'statistics'.

6.2.3 Inference, estimation and confidence

The purpose of obtaining a sample and describing it in detail is to extrapolate the patterns or relationships which are found in the sample on to the population. This process is termed 'inference'. Menges (1973: 3) suggests the following rather more formal definition: 'inference is the reduction of uncertainty by means of statistical experiment and observation'. The uncertainty arises because we do not know about the population in any detail, our only information being about the sample. By applying a series of statistical procedures, this uncertainty may be reduced, but not eliminated. A typical example is the calculation of the population mean. As this is generally not known, it is common to 'estimate' it using sample data. The use of the term 'estimate' implies that the sample value may not be accurate, indeed, it may be wildly inaccurate. Unfortunately, without a full enumeration of the population it is not possible to be certain that the estimate is accurate. However, this does not mean that it is impossible to assess its accuracy. So long as the sample has been drawn in an appropriate way researchers are usually able to assess the accuracy of the estimate to within certain levels of confidence. The justification for this being that, under certain circumstances, it can be shown that sample information will behave in consistent, reproducible ways with respect to the population; ways which allow general assessments of accuracy to be made.

The quality of inference depends on how adequately the sample represents the population. If the sample is some sort of microcosm, a population in miniature, inference is likely to be reasonably accurate. If, however, the sample is wholly arbitrary, or has been gathered together without respect for known features of the population, inference is likely to be of limited value. However, having said this, it is important to realise that the sample cannot fully reproduce the characteristics of the population. This is not altogether surprising given the difference in size between the two, the limitation of the sample to chosen locations, and the possibility that the sample will possess wholly idiosyncratic information. At best, all that the researchers may hope for is that the inferences are as good as may be obtained under the circumstances. What factors help the researchers to assess the accuracy of these inferences?

6.3 OBTAINING SAMPLES AND ASSESSING ACCURACY

6.3.1 Defining the population to be sampled

The first factor which helps with this task is to ensure that the sample is as representative of the population as is possible. This means it should exhibit the key features of the population which are already known to the researchers. However, in order to assess this, the researchers clearly need to have access to some information on the population, for example, on its overall magnitude and the characteristics of its members. Unfortunately, this sort of information is rarely available as populations are usually too large and extensive to document in their entirety. The task of creating a full list of population members, if one does not exist already, is also rarely feasible, as the effort and expense involved can be considerable. For example, the 1981 Census required five years of detailed planning, an administrative staff in excess of 100,000 people and a budget of nearly £50 million in order to generate the full listing of the national population. This sort of expenditure and effort limits the frequency of the exercise to once every decade, suggesting that for any given time other than Census night, the characteristics of the population are unlikely to be understood in full detail. The practical issue is clear: inference operates in the face of uncertainty, both over the state of the sample, and of the population it is hopefully going to represent.

There are a number of ways of partially rectifying the problem of knowing very little about the population which is to be sampled. The first involves the use of so-called 'biased populations'. These are clearly-defined subgroups of the population which have been enumerated already and described in some detail. For example, information collected for a study of students could be used as proxy data for a more general study of British youth; similarly, voters in Cheltenham could be used for a study of the British electorate. The advantage of a biased population is that it provides some information about the population, admittedly biased towards specific subgroups, when none is otherwise available. As a result, the researchers may be able to produce some form of inferential analysis by using the information from the biased population to calibrate the sample. The obvious disadvantage associated with this approach is that the degree of bias involved may be so great that few inferences of quality may be made between the sample and the real population.

A second approach which may be of help if there is no list of the population is to adopt some form of multi-stage 'area' procedure in which the population is subdivided in successive stages into increasingly smaller groupings. The idea behind this approach is that detailed information need only be obtained for these groupings of the population rather than for the whole. Such an approach was adopted in the Cardiff Consumer Panel Survey (Guy *et al.* 1983; Wrigley *et al.* 1985) in order to identify study areas for panel recruitment.

The population in this research project consisted of households located throughout the city of Cardiff, a population for which no complete enumeration of members already existed. The following steps were employed to select a suitable sample which could represent this population but which would not require the researchers to obtain information for households in the whole city. First, Cardiff was subdivided into convenient areal units (enumeration districts). Second, these were classified and grouped into eight sets according to the levels of three binary variables:

1 accessibility to local shops (divided into 'good' or 'poor' if there were more (less) than eight food shops within a quarter mile of the centroid of the enumeration district);
2 accessibility to district centres (divided into 'good' ('poor') if a district shopping centre was located nearer (farther) than a half mile from the enumeration district centroid); and
3 potential mobility (divided into 'good' ('poor') if the car-ownership rate in the enumeration district was above (below) 0.5 cars per household).

These variables were used because it was known that they exert important influences on the patterns of retail behaviour. Third, having categorised and classified the enumeration districts, the next step in the search for a suitable sample was to select at random a single enumeration district from each of these groupings. Fourth, having identified these, lists of electors living in each area were identified from the appropriate electoral rolls for the city. These provided area populations of many hundreds of electors who were then sampled systematically to yield the areal samples. The main advantage of this type of approach which narrows the area for study from the whole of the city to eight enumeration districts, is that it greatly reduces the task of obtaining documentary information on the population. This is because such information was only required for the eight representative areas rather than for the city as a whole. Once this was available, it was relatively easy to select a subset of households from the lists of electors living in each area to form the final sample for the city-wide analysis. (For further details of the procedures involved, or to see how to convert electoral lists into household lists, see Guy *et al.* 1983: 26–9.)

6.3.2 Sampling procedures

The second factor which helps in sample assessment concerns the selection of the sample from the population or biased population lists. A variety of different procedures exist, but only some of them allow researchers to assess the accuracy of estimates. A distinction may be drawn between non-statistical or 'purposive' approaches in which samples are selected on the basis of the professional experiences or intuition of the researchers, and statistical approaches which involve the use of some form of independent random selection. Detailed descriptions of both are given in, among others, Dixon and Leach (1977), Moser and Kalton (1971) and Blalock (1979). Only samples selected by the statistical route may be used in statistical inference.

Within the statistical route there are a variety of different strategies available which may be used to great effect depending on the nature of the problem. These include:

Simple random sampling: a procedure in which every observation is selected at random, possibly using a table of random numbers such as Table 6.1, in which every member of the population, and every combination of members, has a known chance of being selected.

Systematic sampling: a sampling scheme in which every member of the population is numbered in sequence from 1 to *n*, and an observation selected at an equal interval down the sequence. The first observation is usually selected at random, thereafter every fifth, or tenth etc., is selected to complete the sample (this was the procedure used to obtain the samples of households in the Cardiff analysis).

Stratified sampling: a scheme in which the population is first divided into subgroups (strata) on the basis of some key characteristic before being subjected to random or systematic sampling within strata (e.g., as in the selection of areas in the Cardiff example).

Cluster sampling: a scheme popular in geography in which an area is first subdivided into groups and then scanned to find the cluster of groups most like the population. Unlike stratified sampling, in which the strata are defined to be as similar as possible, the clusters are defined to be as different as possible. Once identified, every member of the cluster is included in the sample.

Table 6.1 Table of random numbers

41283	54853	59623	62864	71513
38703	50707	23681	28749	69994
76710	82734	97960	98693	69675
94796	48944	21155	84956	73133
22585	96329	29146	92020	14572
61580	90412	54439	70275	47853
13319	59719	29186	20897	18455
12577	47894	69819	11090	92354
68118	45138	13085	00483	24365
95383	02393	18059	10309	82285
42175	13422	30328	72270	42746
50094	68148	69384	83397	71073
15635	07476	53531	85104	34346
82391	19052	06610	50066	86236
77339	37824	82688	27795	94339
73429	77942	70852	08563	12837
87839	01573	87328	67317	07684
55467	28497	60966	83378	26170
71712	02955	43433	89843	46948
85204	19917	52237	53272	11909
95215	61662	13510	53802	51806
41001	53137	58587	45837	23606
90746	00066	04531	88028	22673
29619	82885	73885	07067	23688
11483	59903	11115	65396	82186

Each of these four major statistical sampling schemes provides researchers with a means of assessing the accuracy of sample statistics. However, they work in different ways and the costs and benefits of each vary.

Simple random sampling, when applied correctly to an appropriate population, provides the most effective way of obtaining information on population characteristics. It is the most basic of the four schemes and is frequently assumed as the sampling scheme underlying the development of inferential methods (Kish and Frankel 1974). Unfortunately, it may also be an expensive and unduly tedious strategy to implement, and may yield samples which are impossible to employ. This latter problem arises because populations which are appropriate for processing by simple random sampling should generally be uniformly or evenly distributed in space or time. To illustrate this, consider the problem of applying simple random sampling to a geographical population which is patently not uniformly distributed: the settlement pattern in a country. Keyfitz (1945) illustrates the problem using Canadian data. In order to obtain a sample of areas which could be used to assess the size of the Canadian population, the country was divided into 30,000 grid squares, each of which covered about 100 square miles on the ground. A series of 1 per cent samples of these was then selected using simple random sampling. The estimates of Canada's population which were produced varied from less than one million to more than fifty million depending on which grid squares were selected for the samples.

The main problem exhibited here is that simple random sampling does not provide sufficient control over the composition of the sample. Not only does it allow the sample to produce highly erroneous estimates of population parameters, it may also allow observations to be included which are distributed so far from each other as to be unusable in practice, for example, villages located across the width of a subcontinent.

The alternative strategies differ from simple random sampling in that they attempt to improve both the efficiency of the data collection process and offer greater control over sample composition. The latter pair, stratified and cluster sampling, are particularly valuable when the population is not uniform as they can allow the researchers greater control over sample composition. Indeed, for many types of geographical problem, some form of stratification or cluster sampling should usually be sufficient to control sample composition. However, Kish (1967) notes that there are always situations in which the need to control which members enter the sample can outstrip the facilities provided by either stratification or clustering. Faced with this, many researchers tend to abandon statistical procedures altogether in favour of purposive sampling or sampling based on the use of some form of underlying model. This retreat from probability sampling may, however, be unnecessary, as a controlled form of simple random sampling does exist which goes beyond what is available from either stratification or clustering. This approach is based on the work of Avadhani and Sukhatme (1965, 1968, 1973), and makes use of results developed in the experimental design literatures. O'Brien (1987a) provides a geographical illustration of the approach using retailing data.

6.4 SAMPLING DISTRIBUTIONS

Having obtained a sample using a statistical sampling strategy, researchers may begin to use it to infer the characteristics of the population. In practice it is usual to select only a single sample rather than several. Given that this single sample may be largely idiosyncratic, how then do the researchers know that the statistics generated from it are accurate of population parameters? Some important theoretical results obtained over many years by experiment and observation provide a possible clue. These concern the sampling distributions of each of the sample statistics.

6.4.1 Sampling distribution of the mean

The simplest sampling distribution to consider is the sampling distribution for the mean. Imagine that a collection of equally-sized samples has been selected from a table of random numbers using simple random sampling (Table 6.2). (It is possible to generate such a table in MINITAB using the RANDOM command with the UNIFORM subcommand.) As the random numbers table used is common to all samples, we can say that the samples are drawn from the same population. The means of each of these samples are different because they reflect different sequences of observations. However, as the means of the sixty samples of size 10 in Table 6.2 show, they may not differ from each other by very much. A graph of these sixty

Table 6.2 Means from randomly-generated samples

(a) 60 sample means ($n=10$)

4.325	5.561	4.841	6.026	3.335	4.261	4.942	5.310	4.983
3.607	5.719	4.052	3.740	5.532	4.841	5.188	5.718	3.367
5.105	5.654	4.725	6.249	4.704	3.417	5.067	3.652	4.362
2.977	5.345	5.792	5.590	5.010	3.007	5.635	2.938	5.255
4.887	4.031	5.199	4.184	4.160	4.729	5.954	3.756	3.555
4.177	4.555	4.751	4.630	4.341	3.557	5.396	4.980	3.213
5.774	4.050	4.520	3.610	4.068	4.449			

(b) 60 sample means ($n=100$)

4.255	4.518	4.367	3.907	4.375	4.640	4.619	4.505	4.454
4.847	4.295	4.240	4.565	4.790	4.093	4.407	3.997	4.296
4.492	4.238	4.622	4.502	4.197	4.521	5.000	4.602	5.007
4.411	4.686	4.619	4.498	4.293	4.419	4.636	4.522	4.744
4.373	4.776	4.575	4.749	4.441	4.157	4.567	4.598	4.146
4.432	4.583	4.263	4.485	4.778	4.053	4.207	4.441	4.498
4.362	4.815	4.660	4.479	4.524	3.869			

(c) 60 sample means ($n=200$)

4.106	4.553	4.594	4.568	4.642	4.320	4.213	4.507	4.400
4.699	4.509	4.509	4.465	4.766	4.301	4.436	4.608	4.624
4.367	4.558	4.364	4.711	4.377	4.243	4.798	4.331	4.527
4.603	4.411	4.384	4.712	4.443	4.391	4.357	4.303	4.332
4.558	4.821	4.580	4.270	4.429	4.496	4.309	4.543	4.534
4.370	4.273	4.532	4.701	4.438	4.578	4.141	4.465	4.544
4.312	4.271	4.506	4.451	4.604	4.806			

Notes: Overall means: sample (a) – 4.6060
sample (b) – 4.4668
sample (c) – 4.4761

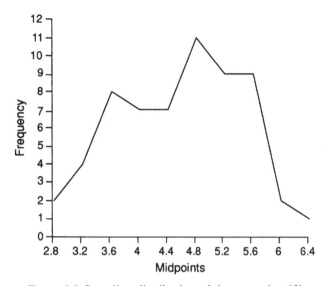

Figure 6.1 Sampling distribution of the mean ($n=10$)

means is given in Figure 6.1. This displays the sampling distribution of the mean based on the generated data.

The similarity of the sixty means suggests that the population mean is likely to be somewhere within the range 3 to 6. This evidence is, however, only indirect and is based solely on the findings of sixty equally-sized random samples drawn using the same sampling strategy. As it stands, the evidence for the value of the parameter is not particularly strong and it is not clear whether similar findings would have occurred if different samples were used, for example, a similar number of larger samples. The sampling distributions for sixty samples of size 100 and size 200 are given in Figures 6.2 and 6.3. These indicate an important and interesting feature of the sampling distribution of the mean, namely, that it tends to become more Normally distributed as the size of the sample gets larger. This tendency is true even if the original sample readings are not Normally distributed, as can be seen from the raw observations in Table 6.2. Moreover, it can be shown that the mean of the sampling distribution of means will equal the population mean if the sample size is sufficiently large.

This latter comment is the key factor in assessing the accuracy of a sample mean. What it actually states is that the sample mean will on average give the correct value for the population mean and that the shape of the sampling distribution of the mean will always be approximately Normal. These characteristics are based on the Central Limit Theorem properties which were outlined in Chapter 5, and allow researchers therefore to limit their sampling to a single sample, because they know

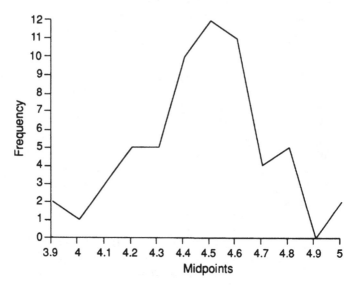

Figure 6.2 Sampling distribution of the mean ($n=100$)

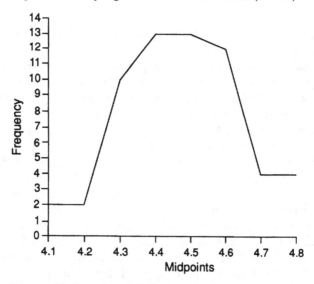

Figure 6.3 Sampling distribution of the mean ($n=200$)

that on average the sample mean and the population mean will be equal. However, this is not a universal property, it only applies if the conditions underlying the Central Limit Theorem are met by the sample data. These are:

1 the original observations must be irregularly distributed;
2 the sample must be 'sufficiently large';

132

3 the observations must be independent (that is, the presence of a specific observation in the sample does not prejudice the chances of any other observation being selected).

These conditions are vital: if they do not apply then the tendency for the sampling distribution of the mean to approach Normality as suggested by the Central Limit Theorem will not be exhibited.

This may be illustrated using the data in Table 6.3 which were collected for a study of dewfall in Oman (Anderson and O'Brien 1988). The observations used in the sampling (dewfall measurements) were collected sequentially over eleven days at a number of different sampling sites in six locations. The sampling distribution of the mean for samples of size 10 are shown in Figure 6.4 and for samples of size 65 in Figure 6.5. In spite of the larger sample size the tendency to approach Normality is not exhibited, principally because the observations are correlated with each other both spatially and temporally and were not randomly and independently selected. Thus, though it is possible to add them together to produce a mean value, and to see that they are irregularly distributed, a key condition of the property (random, independent selection) has not been met.

Table 6.3 Dewfall data from Oman (extract)

Sensor	Sample days							
	1	*2*	*3*	*4*	*5*	*6*	*7*	*8*
1	0.573	0.893	0.000	0.253	0.947	1.598	0.000	0.069
2	1.318	1.241	0.609	0.000	0.871	2.105	0.016	0.005
3	0.224	0.676	0.215	0.074	0.408	0.930	0.000	0.019
4	1.030	1.141	0.572	0.372	1.140	1.505	0.068	0.127
5	0.000	0.898	0.061	0.210	1.164	1.707	0.000	0.020
6	0.318	0.364	0.097	0.017	0.242	0.353	0.009	0.002
7	0.759	0.777	0.629	0.400	1.210	2.015	0.127	0.085
8	0.890	0.022	0.534	0.523	1.065	1.577	0.428	0.127
9	0.971	1.303	0.670	0.126	1.107	2.115	0.000	0.029
10	0.889	1.118	0.096	0.233	1.134	1.724	0.065	0.055

Random sampling is assumed by the property because it ensures that the observations are selected independently, that they can be added together (as they are to produce the mean), and that they can range widely and irregularly in value. If these conditions can be met, the tendency will emerge empirically even for samples drawn from non-Normal populations. However, the number of observations required in the sample before the tendency emerges depends on the underlying shape of the population. In general, samples drawn from non-Normal populations need to be considerably larger than those drawn from Normal populations. Ehrenberg (1982) suggests that at least one hundred observations need to be included in samples drawn from non-Normal populations whereas samples of thirty

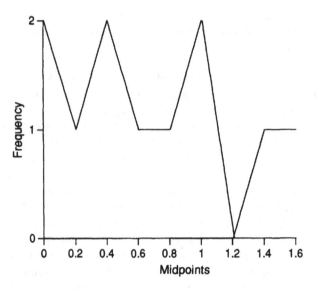

Figure 6.4 Omani data (*n*=10)

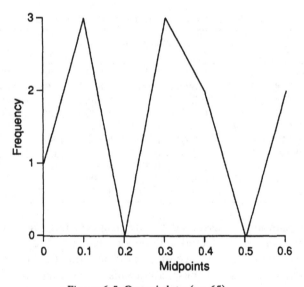

Figure 6.5 Omani data (*n*=65)

or more drawn from Normal populations are usually sufficiently large. However, he also notes that the tendency begins to be apparent in samples drawn from Normal populations which include only ten observations.

Whilst the shape of the sampling distribution of the mean relies on established theories, it may be useful to illustrate the link between the average sample mean and the population mean with an example. Consider the following small population consisting of three numbers:

1 2 3

The mean of these (the population mean μ) is 2. Nine separate samples containing two observations may be drawn from this population. These, with their respective sample means, are:

Combination of observations	Sample mean
11	1
12	1.5
13	2
21	1.5
22	2
23	2.5
31	2
32	2.5
33	3
Total	18

The mean of these nine samples is $18/9 = 2$, which is identical on average to the population mean. The histogram of this sampling distribution is given in Figure 6.6.

In the above example it is possible to extract nine different samples of size 2 from the small population if sampling takes place with replacement. This means that each observation is available for inclusion more than once in each sample. The idea of replacement can be illustrated if

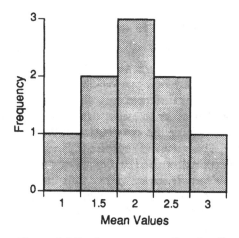

Figure 6.6 Replacement sampling ($n=9$)

one considers the small population as consisting of three coloured balls in a bag, for example 1 = white, 2 = blue, 3 = red. A single ball is selected from the bag and its colour noted. It is then replaced in the bag and a second ball drawn out. As the first ball has now been replaced it is perfectly possible for it to be drawn a second time to produce the second half of each two-ball sample. This is how it is possible to include combinations such as 11, 22 and 33 in the nine samples listed above. However, sampling may also take place without replacement. In this case once a ball has been drawn from the bag it is not replaced and so cannot be selected a second time: combinations such as 11, 22 and 33 are thus impossible. The main effect of this is that only three samples may be drawn from the small population if non-replacement sampling is used:

1 2
2 3
1 3

However, if the means of these samples are calculated (1.5, 2.5 and 2), and thereafter the mean of the means (2), it is immediately clear that the restricted sampling has not affected the tendency of the mean of the sampling distribution to equal the population mean on average. Because of this property, sampling without replacement is frequently used in social survey work. The need to process fewer observations may, in turn, make the process of sampling faster without introducing any serious errors. This does not, however, mean that non-replacement sampling can always be used. Systematic errors may be introduced into the analysis by a non-replacement strategy if the size of the sample relative to the population is large. The main effect of this may be noticed in the calculation of the variance and standard deviation measures, rather than in the calculation of the mean.

6.4.2 The standard error

The fact that the sampling distribution of the mean will on average be identical to the population mean is an important and interesting empirical finding (that is, established by observation or experiment). However, it does not tell us how far, or different, any single sample mean will be from the population mean. In order to assess this, it is necessary to calculate a summary measure describing the distribution of the sample means around their average value. This measure represents the same sort of information as the standard deviation described earlier, except that in this case, it refers to the distribution of sample means rather than individual observations in a single sample. The term used to refer to this measure is the standard error (or alternatively, the standard error of the mean). It is interpreted as showing the average error of an observed sample mean from the population mean.

The standard error may be calculated in a number of different ways, either empirically, by researchers generating all possible samples of size n from the population, calculating the mean of their sampling distribution, then calculating the distribution of the individual sample means around this, or by a theoretical short cut. The former, if adopted, would be tedious and require the researchers to undertake a task for which they do not have any information: data from all possible samples of size n. As, in practice, only one sample is obtained, the alternative strategy is usually adopted. This procedure makes use of the fact that the standard error may be shown theoretically to be equal to the ratio of the population standard deviation to the square root of the sample size:

$$\frac{\sigma}{\sqrt{n}} \tag{6.1}$$

In other words, it is possible to use the scatter of the individual readings in the population to calculate the standard error rather than use the sample means from all possible samples. However, there is still a major problem: the population standard deviation is itself usually unknown. Because of this, it is usual to replace equation 6.1, which provides an accurate calculation of the standard error, by an alternative formula (equation 6.2) which estimates it:

$$\frac{s}{\sqrt{n}} \tag{6.2}$$

In this refined version, the population standard deviation is replaced by the sample standard deviation. This may seem a little curious because it means that the assessment of the accuracy of the single sample mean from the unknown population mean is provided by calculating a measure based solely on the sample data, that is, on the ratio of the observed standard deviation from the single sample of data to the square root of the number of observations in it. This may seem rather strange and arbitrary. Why does it work?

The mechanism which allows this procedure to work is based on the distributional properties of the sampling distribution of the mean. As we have seen, if certain conditions hold, the distribution of the sample means tends towards normality and the mean of the sampling distribution equals the population mean on average. Because of this tendency, it is possible to use the attractive distributional properties of the Normal distribution to describe the relationship of a single sample mean with the population mean. The main characteristics of the Normal distribution were set out in Chapter 5 but may be repeated here. If the distribution of sample means is roughly Normal then:

1 68 per cent of all possible sample means will lie within plus or minus 1 standard error of the population mean.

2 95 per cent of all possible sample means will lie within plus or minus 2 standard errors of the population mean.

Whenever the sample is large enough (greater than about 10 for samples drawn from Normal populations, or 100 for samples drawn from non-Normal populations) the same distributional properties will emerge for the estimated standard error. In other words:

1 About 68 per cent of all possible sample means will lie within plus or minus 1 estimated standard error of the population mean.
2 About 95 per cent of all possible sample means will lie within plus or minus 2 estimated standard errors of the population mean.

This means that the fact that the observed sample data are being used to provide the link between the sample and the population does not make the assessment invalid as the distributional properties under σ and s are similar (i.e., the differences between them are comparatively trivial).

6.4.3 Small samples

There is a slight complication to this relationship which becomes evident in small samples. In this context, 'small' refers to any sample from a Normal (non-Normal) population which contains less than about 10 (100) observations. Below this threshold, the Normal distribution does not provide an adequate description of the distribution of the sample means, and so should not be used. Instead, the t-statistic (alternatively termed the t-ratio or t-variable) provides a viable alternative. This is described algebraically as:

$$t = \frac{\bar{x} - \mu}{(s / \sqrt{n})} \tag{6.3}$$

where:

$\bar{x}$ is the observed sample mean
μ is the unobserved population mean
s is the calculated sample standard deviation
$\sqrt{n}$ is the square root of the sample size

(Note: the denominator of equation 6.3 is the formula for the estimated standard error.) This statistic is valuable because its theoretical distribution is known to follow the t-distribution (see Chapter 5). This distribution gives the proportion of times different values of the t-statistic are likely to occur for all possible samples of a given size from a Normal or near-Normal population. Its shape depends on two things:

1 the size of the sample (n), and
2 the number of degrees of freedom available.

The latter condition reflects the number of independent pieces of information existing in a data set.

To illustrate the idea of independence, consider the following example in which five observations are added, their average calculated, and the difference in value between each observation and the average displayed:

Observations:	1	2	3	4	5
Sum:	15				
Average:	3				
Differences:	−2	−1	0	1	2

The sum of the differences equals 0. This should always be the case when differences between individual observations and their mean are calculated and summed. Because of this fact, once the mean and four of the differences are known it is possible to calculate the fifth automatically. The sum of the first four differences is −2, therefore, to satisfy the condition that the sum of the differences must equal 0, the fifth difference must equal 2, as indeed it does. The value of this difference is therefore dependent on the other values rather than independent once the mean has been calculated. This leads to a general condition which is that the calculation of each statistic (mean, variance, etc.) reduces the number of independent items of information available within the data by one. Readers should note that independence and dependence refer to the information content of the data and not to specific observations. It therefore is not correct to assume that observations 1 to 4 are independent and observation 5 is dependent. If a data set contains five observations, and a mean value has been calculated, there are four independent pieces of information left, because one is lost in the calculation of the mean.

The shape of the t-distribution depends on the number of degrees of freedom in the data (Figure 6.7). This shape may also be summarised numerically as in Table 6.4. The figures in the table show the limits either side of the population mean which contain varying percentages of the t-statistic. Notice that the value of these limits depends on the degrees of freedom involved. The figures in the table may be interpreted either in terms of t or the sample mean. In the former, for 90 per cent (95 per cent, 99 per cent) of all possible random samples containing 21 observations (20 degrees of freedom) the value of the t-statistic will lie within plus or minus 1.7 (2.1, 3.9). In the latter, it implies that 90 per cent (95 per cent, 99 per cent) of all possible sample means lie between plus or minus 1.7 (2.1, 2.9) estimated standard errors of the population mean. By extension, it follows that 10 per cent (5 per cent, 0.1 per cent) of all possible sample means lie outside plus or minus 1.7 (2.1, 2.9) estimated standard errors of the population mean. The t-distribution thus provides a mechanism for investigating the extremities of the distribution of sample means.

Table 6.4 Values of *t*

Degrees of freedom	.2 / .1	.1 / .05	Significance .05 / .025	.02 / .01	.01 / .005	.002 (2TT) / .001 (1TT)
1	3.08	6.31	12.71	31.82	63.66	318.31
2	1.89	2.90	4.30	6.96	9.93	22.33
3	1.64	2.35	3.18	4.54	5.84	10.22
4	1.53	2.13	2.78	3.75	4.60	7.17
5	1.47	2.02	2.57	3.37	4.03	5.89
10	1.37	1.81	2.23	2.76	3.17	4.14
20	1.33	1.73	2.09	2.53	2.85	3.55
60	1.30	1.67	2.00	2.39	2.66	3.23
120	1.29	1.66	1.98	2.36	2.62	3.16
infinity	1.28	1.65	1.96	2.33	2.58	3.09

Note: 1TT = one tailed test
2TT = two tailed test

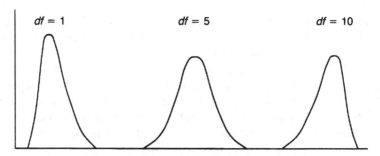

Figure 6.7 Various *t*-distributions

As the distributions in Figure 6.7 show, the values of *t* are greatest when the degrees of freedom are small and decrease to a limit as they get larger. Above 30 the values of *t* hardly change. In fact, the figures in the final row are identical to those produced by the Normal distribution. This confirms the fact that as the sample size increases the *t*-distribution approximates a Normal distribution.

6.4.4 Summary

It may be useful at this point to state the main findings of the last three subsections as they are of considerable practical significance for what follows. The main point being made is that it is possible to use summary information from a single observed sample to describe characteristics of an unobserved population to within calculable levels of accuracy. Because of the tendency for the sampling distribution of the mean to approximate a Normal distribution, it is possible to use the Normal curve to describe the relationship of a single sample mean to the unobserved population mean. This tendency works for samples which have been obtained from both Normal and non-Normal populations but, in the latter case, the sample needs to be considerably larger. However, it rests on the applicability of a series of underlying conditions, particularly the need for the independent selection of observations. If these do not hold, it will not emerge, and the most straightforward and easily adaptable inferential results cannot be applied.

The behaviour of the sampling distribution of the mean is unusual in that it may be described with relative ease using the Normal curve. This is not the case with the sampling distributions of other summary measures (for example, the variance), or with sampling strategies other than simple random sampling. Though Ehrenberg (1982: 118) notes that most sampling distributions, including those of the variance and other measures, tend to be approximately Normal if the sample size is sufficiently large, for some this means that samples might need to contain at least 1,000 observations. This is a larger number of observations than is frequently collected for geographical samples, suggesting that the Normal curve will not be able to assess their ability to stand as representatives of the population. As a result, other theoretical distributions (t, F and chi-square) may have to be used instead of the Normal. Chatfield (1983), for example, notes that a chi-square distribution may be used to describe the sampling distribution of the variance.

6.5 ESTIMATION AND CONFIDENCE

6.5.1 Estimates and estimators

The preceding section has shown that it is possible under certain circumstances to use the information in a single sample of observed data to represent some of the characteristics of an unseen population. The results presented so far show how it is possible to relate sample measures such as the mean to their population counterparts using empirically-derived relationships from sampling distributions. However, the fact that one can say that 95 per cent of all possible sample means, generated by simple random sampling, will be within plus or minus two estimated standard errors of the population mean, does not state what that population mean actually is. To do this, one has to provide an estimate of the population mean from the sample data.

This section presents some information on the processes of estimating population parameters from sample statistics. The main points to be covered are:

1 Estimation is a two-stage process involving, first, the calculation of a numerical value for the parameter (a 'point' estimate), and thereafter,

assessing its accuracy by establishing a level of confidence in it (an 'interval' estimate).

2 There are a variety of alternative ways of calculating an estimate. As a result, it is usual to talk of different types of estimator of the parameter, to reflect the fact that different formulae and logic may be used in its calculation.

3 The estimation of parameters under simple random sampling is usually assumed in the literature. This does not preclude the use of alternative forms of statistical sampling, but the formulae required to describe these are generally more complex (for a detailed discussion of this issue, see Kish and Frankel 1974).

The simplest parameter to estimate, regardless of which estimation procedure is to be used, is the population mean. In the absence of other information, or information to the contrary, it is usually advisable to make use of sample statistics as estimators of parameters. Thus, the most widely-used estimator of the population mean is the sample mean; the justification being that, on average, the sample mean and the population mean are identical. (Though this is generally taken to be the case, Chatfield and Collins (1980: 103) point out the existence of so-called James–Stein estimators which, when applied to artificially-constructed or otherwise 'special' data, appear to be even more appropriate.)

The estimation of other summary parameters is rather less straight-forward. The standard formula for the variance:

$$\text{variance} = \frac{\text{sum of squared deviations from mean}}{\text{sample size}}$$

or

$$\sigma^2 = \frac{\Sigma (x - \bar{x})^2}{n} \tag{6.4}$$

cannot be used as it stands because its average value across all possible samples is smaller than the population variance. This may be illustrated using the small population of three numbers presented in the previous section. In this population, the members are:

1 2 3

and the mean and variance are 2 and 0.67 respectively. The estimates of the variance derived from the nine samples using replacement sampling are:

Combinations:	11	12	13	21	22	23	31	32	33
Variance:	0	.25	1	.25	0	.25	1	.25	0
Average variance:	0.34								

The average variance produced by sampling is thus lower than that calculated for the population. However, if a modified formula is used:

$$\text{variance} = \frac{\text{sum of squared deviations from mean}}{\text{sample size minus one}}$$

or

$$\sigma^2 = \frac{\Sigma(x - \bar{x})^2}{n - 1} \tag{6.5}$$

then revised estimates of the population value are produced:

Combinations:	11	12	13 21	22 23	31 32	33
Variance:	0	.5	2 .5	0 .5	2 .5	0
Average variance:	0.67					

The average of these is identical to the population value. (Readers might like to estimate the variance under non-replacement sampling to see if it, too, produces a value identical to that calculated for the population.)

The sample estimator of the population standard deviation can be used in a similar way. However, the formula:

$$\sqrt{\frac{\Sigma(x - \bar{x})^2}{n - 1}}$$

yields an incorrect answer for the population measure. The reason for this lies in the fact that the average of the square root of a set of numbers is not equal to the square root of the average. This can be seen from the example. The variance estimates under the modified estimator are:

0.0   0.5   2.0   0.5   0.0   0.5   2.0   0.5   0.0

which yield an average variance of 0.67. The equivalent standard deviations (square roots of the variance) are:

0.0   0.7   1.4   0.7   0.0   0.7   1.4   0.7   0.0

which yield an average of 0.62. This value is different from either the population value or the square root of the average variance. In spite of this, the formula for the sample standard deviation still tends to be used as the estimator of the population standard deviation. Ehrenberg (1982: 124) lists three reasons for this:

1 The degree of error introduced by the use of the standard formula is usually small (especially if the sample size is small relative to the population).
2 The requirement that the estimator be error-free is not absolutely essential.
3 There is no unambiguously better alternative.

The third of these reasons amply illustrates the fact that the choice of an estimator tends to be a matter of some experimentation and compromise rather than a cast-iron procedure.

6.5.2 'Best' estimators

For any parameter there are likely to be several alternative estimators available. For example, in assessing central tendency, the researchers could use the arithmetic mean, geometric mean, median or the average of the range. The question which arises, therefore, is which of the alternatives to use. It seems reasonable to assume that the estimator which is best should be used. However, 'best' is a subjective term and requires a rather more formal specification before it may be applied operationally.

The following are generally considered desirable properties in an acceptable estimator (after Silvey 1975; Hanuschek and Jackson 1977; Chatfield 1983):

1 *Lack of bias* – the estimator should give the correct value for the parameter on average across all possible samples.
2 *Efficiency* – the variance associated with the estimator should be as small as is possible.
3 *Consistency* – the estimator should approach the value of the population parameter as the sample size increases (more formally, approaches in probability the true value of the parameter).

As was seen previously, the sample mean and variance satisfy requirements 1 and 2, and it may be shown from theory that 3 also applies to them. Unfortunately, however, as Menges (1973: 11) notes, 'these criteria, taken individually, are incomplete, and, to a certain extent, contradictory'. They are contradictory in the sense that, on occasion, it may prove to be impossible to satisfy all three properties simultaneously.

Two problems are frequently met. First, an estimator which appears to be satisfactory according to one criterion may not be so according to others. Second, an estimator which appears to be acceptable when applied to large samples may fail to be so in small samples (McFadden 1976; Shenton and Bowman 1977). This is a problem of considerable importance and is a topic of current statistical research. However, in spite of this potential dilemma, it seems to be generally accepted that an estimator must exhibit minimum variance and unbiasedness to be considered 'best'.

6.5.3 Moments estimators

There is a variety of procedures available which may be used to produce acceptable estimates. The two most widely used are based on general estimation strategies:

1 The method of moments.
2 The method of maximum likelihood estimation.

The moments are essentially functions of the sampling distributions of a random variable. They may be defined with respect to the origin or to the mean. Hanuschek and Jackson (1977: 328) note that the following general formulae may be used to calculate the moments about the origin of discrete and continuous random variables. For discrete variable, X, this is:

mth moment $= \Sigma x_i^m f(x_i)$

where

x_i is the ith outcome associated with the discrete variable, and
$f(x_i)$ is its associated probability of occurrence

The formula for continuous variable, X, is rather more complex:

$$m\text{th moment} = \int_{-\infty}^{\infty} x^m f(x) \, dx$$

where

$\int$ refers to an unbounded integral, that is, allowing integration over the range negative to positive infinity
x, f are as for the discrete case

For the discrete data in Table 6.5 the first moment is calculated to be 83.583 (or 83,583 because the raw data are measured in thousands). Notice that this is equivalent to the mean value as calculated for grouped data

Table 6.5 Illustration of moments estimation (discrete data)

Midpoint	Frequency	Probability	Product
80	1	.08333	6.66666
81	2	.16666	13.49999
82	4	.33333	27.33333
84	1	.08333	6.99999
86	1	.08333	7.16666
87	1	.08333	7.24999
88	2	.16666	14.66666
Sums	12	1.0	83.58328

Notes:
1 Data are taken from Table 4.5 (East Anglian unemployment)
2 Midpoint refers to midpoint of class
3 Probability values generated by dividing class frequency by 12 (the size of the data set)
4 Product refers to the product of the midpoint values multiplied by their respective probabilities

(see Chapter 4, section 4.4.3). This suggests that the mean of a series of observations is equivalent to the first moment about the origin. Higher-order moments about the mean correspond to standard summary measures such as the variance, skewness and kurtosis.

The basic idea of moments estimation may be extended to provide estimates for the terms in linear models which are unbiased and have minimum variance. In Chapter 5, Section 5.2, a selection of these simple models was presented and an estimate of the slope coefficient obtained directly from the data. To illustrate the process of moments estimation as applied to this problem, consider once again, the simple six-times table. From experience, we know that the slope coefficient relating X and Y is 6. However, how might this be estimated from the raw data?

A procedure which may be used to estimate linear relationships is ordinary least squares estimation. This is a procedure based on the first two moments of a distribution, that is, on the use of mean and variance information. In order to apply this procedure, the basic equation (equation 6.6) may be modified slightly to incorporate a term which represents errors – items not incorporated explicitly in the model (equation 6.7):

$$Y = \alpha + \beta X \tag{6.6}$$

$$Y = \alpha + \beta X + \varepsilon \tag{6.7}$$

In these equations,

    Y represents the response component
    X represents the explanatory or controller component
    α is the population intercept or constant parameter
    β is the population slope parameter
    ε represents the error component

Equation 6.7 relates to the population model which, like most populations, is usually approximated by an appropriate sample. The equivalent sample model to equation 6.7 for $i = 1, n$ observations is:

$$y_i = a + bx_i + e_i \tag{6.8}$$

where,

    y_i is the ith sample observation on Y
    x_i is the sample observation on X
    a is the sample intercept or constant term
    b is the sample slope term
    e_i represents residuals

The residuals in the sample model are the differences in value between the observed values of $-y_i$ in the data, and the estimated values, $-\hat{y}_i$, given the model. That is:

residuals = observed y − expected y

or

$$e_i = y_i - \hat{y}_i \tag{6.9}$$

(The difference between errors and residuals is discussed in more detail in Chapter 8.)

In least squares estimation, 'best' estimators are produced when the sum of squared residuals from the sample model is minimised:

$$\text{minimise} \quad \Sigma e_i^2 = \Sigma(y_i - \hat{y}_i)^2 \tag{6.10}$$

Minimisation is achieved by differentiating equation 6.8 with respect to a and b. This yields the so-called normal estimating equations (see Pindyck and Rubinfeld 1976 for more details) which, after some manipulation, yield the following two expressions for a and b:

$$b = \frac{\text{sum of cross products of } x \text{ and } y}{\text{sum of } x \text{ squares}}$$

or

$$b = \frac{\Sigma (x_i - \bar{x})(y_i - \bar{y})}{\Sigma (x_i - \bar{x})^2} \tag{6.11}$$

and

$$a = \bar{y} - b\bar{x}$$

The calculations associated with the estimation of the a and b terms for the six-times table data are given in Table 6.6.

In this simple example, the line fits the observed scatter of data points exactly so that there are no residuals to minimise in the first place. Also, it is drawn through the origin ($X=0$, $Y=0$) so that the value of a is 0. If there had been scatter in the observed data and, by implication, in the population in general, this approach to estimation would still have been feasible so long as two assumptions relevant to the behaviour of the errors could be sustained. These are:

1 The errors are independent of each other.
2 The variance of the errors is constant for all values of X.

If either or both of these assumptions are violated by the data, then the simple population model (equation 6.7) may be misspecified and so cannot yield estimators which are BLUE (best linear unbiased estimators). Instead, some modifications may need to be made, either to redefine X in some way, or to use some alternative system of estimation. One alternative which is frequently helpful is generalised least squares (see Pindyck and Rubinfeld 1976 and Kennedy 1979 for brief descriptions of

Table 6.6 Calculation of parameters using least squares: six-times table data

x	$x-\bar{x}$	$(x-\bar{x})^2$	y	$y-\bar{y}$	$(x-\bar{x})(y-\bar{y})$
1	−5.5	30.25	6	−33	181.5
2	−4.5	20.25	12	−27	121.5
3	−3.5	12.25	18	−21	73.5
4	−2.5	6.25	24	−15	37.5
5	−1.5	2.25	30	−9	13.5
6	−0.5	0.25	36	−3	1.5
7	0.5	0.25	42	3	1.5
8	1.5	2.25	48	9	13.5
9	2.5	6.25	54	15	37.5
10	3.5	12.25	60	21	73.5
11	4.5	20.25	66	27	121.5
12	5.5	30.25	72	33	181.5

Sum(x) = 78; Sum(y) = 468; Sum $(x-\bar{x})^2$ = 143; Sum $(x-\bar{x})(y-\bar{y})$ = 858; Mean (x) = 6.5; Mean (y) = 39

$$b = \frac{\text{sum } (x-\bar{x})(y-\bar{y})}{\text{sum } (x-\bar{x})^2} = \frac{858}{143} = 6$$

a = mean(y) − b mean (x) = 39 − 6(6.5) = 39 − 39 = 0

Final model: y=6x

this procedure, and Wald 1943, Neyman 1949 and Bhapkar 1966 for more detailed treatments).

6.5.4 Confidence

Point estimates produce a numerical value for parameters which may or may not be true. The reason for the uncertainty arises because the estimate is based on a sample which may or may not be representative of the population. If it is, then one may expect the sample to produce reasonably accurate estimates of the parameters. However, if the sample is essentially idiosyncratic, the estimate may be very inaccurate. Because only one sample is usually obtained, it is not possible to be absolutely certain about the accuracy of the estimate. However, so long as the sample has been generated in an appropriate way, it is possible to assess its likely error by calculating a confidence interval for it. This provides a range of values within which the population parameter can be expected to be found. If the sample is very accurate of population characteristics, we can expect this interval to be reasonably narrow (for example, we may be able to make statements such as the population mean is 10.5 plus or minus 0.5). However, if the sample is inaccurate, we cannot be so confident in making such statements, because the margin for error is potentially greater.

A confidence interval is created by calculating interval estimates for the population parameter. These are probability statements and correspond to an upper and lower limit for the value of the parameter. The size of this interval reflects the level of confidence associated with the parameter. A 95 per cent confidence interval surrounding an estimate implies that, in the long run, a claim that the population parameter lies in the interval would be correct 95 times out of 100. A 99 per cent interval would increase this accuracy to 99 times out of 100. Thus for a single sample, the confidence level corresponds to the odds (or relative chances) of success-fully containing the parameter in the interval.

The interpretation of a confidence interval is complex, principally because it reflects sampling variability between samples rather than the unknown, but fixed, value of the parameter. If the observed mean were a random variable rather than a single value, then the estimation of specific limits would be true regardless of which sample were drawn. However, as this is not likely to be the case, some care needs to be taken to interpret the limits generated as these relate to the particular sample drawn.

Ehrenberg (1982: 125) notes that this complexity can be intolerable. However, because the means of random samples drawn from the popula-tion are generally similar, it is also likely that the confidence limits will be for any given level of confidence. To illustrate, consider a Normally-distributed data set with mean 10 and standard deviation 0.2. Using the theoretical proofs from the standard Normal curve, we can see that limits of 9.6 and 10.4 correspond to an interval containing 95 per cent of the distribution. For an interval containing 99 per cent of the distribution, the limits would have to be extended to about 9.4 and 10.6. These limits are specific to the sample. A second sample with mean 10.1 and standard deviation 0.25 would generate different limits for any given level of confidence. Notice that in moving from the 95 per cent to the 99 per cent level the chances of correctly containing the population parameter increase disproportionately to the increase in the size of the interval, whatever it actually is. This suggests that if a population parameter lies outside a limit, it is likely to lie just outside.

Because of the regular properties of the Normal curve, the confidence limits can be calculated directly from

$$prob\left(\bar{x} - z\,\frac{\sigma}{\sqrt{n}} < \mu < x + z\frac{\sigma}{\sqrt{n}}\right) = 0.95 \tag{6.12}$$

where

z refers to the value of the z score associated with any specific value of the standard deviation

This relationship will hold if the population variance is known or if the sample is sufficiently large to estimate it accurately. This normally requires the sample to be at least thirty observations, and merely involves replacing σ by its sample equivalent. If the sample is too small for this, the *t*-distribution provides an alternative mechanism. The relationship

$$\bar{x} \pm t\,(s/\sqrt{n}\,) \tag{6.13}$$

provides estimates of the confidence limits for any given level of confidence. This rapidly approaches the values associated with the Normal curve as the sample size gets larger.

6.6 HYPOTHESES: DEVELOPMENT AND TESTING

6.6.1 Some examples of acceptable hypotheses

If a sample may be used to describe the features of a population, it is perhaps not unrealistic to assume that it may also be used to test hypotheses about it. Hypotheses are statements of belief about characteristics of a population which are made in the face of uncertainty. A typical hypothesis might be 'the population mean is 20'. Alternatively, an hypothesis might suggest a range of values: 'the population mean lies between 20 and 25'. In both examples, care has been taken to make the statements as precise as possible. Notice also that the statements apply to the population. The statement, 'the sample mean is 20', though precise, is not considered an hypothesis because it states nothing about the population.

Precision is required so that tests of the hypothesis may be made. However, as the population is usually not observed, these tests are applied to the data gathered in the sample. As sample values will probably differ from those in the population by an amount which merely reflects sampling, it follows that an hypothesis which states 'the population mean is 20' may be supported by a sample whose observed mean is not 20. The hypothesis may be supported if it can be shown that the difference is less than can be expected by chance. If, however, the difference is greater than can be expected by chance, then the hypothesis must be rejected, or a less stringent test needs to be used.

6.6.2 Null and alternative hypotheses

Before introducing some possible tests to clarify this point, it is perhaps useful to set out the terminology and operations which are frequently met in hypothesis testing. The following terms refer to two distinct types of hypothesis:

1 The null hypothesis – the hypothesis which is to be tested.
2 The alternative hypothesis – one of a series of possible alternatives to the null hypothesis.

If we take as a null hypothesis the statement 'the population mean is 20', then the following alternative hypotheses may be suggested:

150

1 The population mean is not 20.
2 The population mean is less than 20.
3 The population mean is greater than 20.

Algebraically, these may be written:

1 H_0: μ $= 20$
2 H_1: μ $\neq 20$
3 H_2: μ < 20
4 H_3: μ > 20.

The notation H_0 is typically used to refer to the null hypothesis, whereas H_1, H_2, and H_3 refer to specific alternative hypotheses. It is important to consider carefully what each of these alternatives implies. For H_1, any difference which is greater than one likely to have arisen by chance will cause the null hypothesis to be rejected. For H_2 and H_3, this difference has to be in a specific direction. Because the direction of difference is specified, tests applied to H_2 and H_3 are termed one-tailed tests. In comparison, for H_1, in which direction is not specified, tests are two-tailed. This terminology will be clarified by examples below and in Part II.

6.6.3 Type 1 and Type 2 errors

The object of hypothesis testing is to find conditions under which a null hypothesis may be rejected. If it can be rejected, it follows that the information it contains is wrong. However, if it cannot be rejected, it does not mean that its information is correct. This is because a more rigorous test may succeed in rejecting it. In general, the rigour of a test reflects the level of significance associated with it. This is a measure of probability associated with the test. A higher level indicates a more rigorous test as it implies that the probability of rejecting the null hypothesis is increased. For example, a value may exceed the 5 per cent level five times in any 100 merely by chance. However, it may only exceed the 1 per cent and 0.1 per cent levels by chance once in every 100, or once in every 1,000, respectively. A null hypothesis value found to exceed these limits is thus very likely to be incorrect.

Unfortunately, there is a dilemma. As the probability of rejecting the null hypothesis increases with the significance level, the chance of rejecting a correct statement increases. This is termed a Type 1 error (see Figure 6.8). However, if the level of significance is reduced to minimise this possibility, the probability of accepting an incorrect null hypothesis increases. This is termed a Type 2 error. The two errors are mirror images of each other, meaning that as one is minimised, the other is maximised. A compromise thus needs to be reached in testing any hypothesis, in which the probabilities associated with both types of error are kept as small as

RESEARCH DECISION

	Accept H_0	Reject H_0
True	Correct result	Type I error
False	Type II error	Correct result

NULL HYPOTHESIS

Figure 6.8 Error table

possible consistent with the needs of the analysis. This simply means that the nature of the null hypothesis tends to determine which type of error, if either, ought to be made. A failure to minimise a Type 1 error, for example, could have disastrous consequences for the users of medicines, engineering constructions (bridges, dams) and aircraft.

Ehrenberg (1982: 137) suggests a few rules of thumb which may help. If the sample result differs from the null hypothesis by less than 1.5 standard errors, then do not reject the null hypothesis as only one in ten samples would produce this difference by chance. However, if the difference is more than 2.5 standard errors, then reject the null hypothesis, as only one in one hundred samples would produce this difference just by chance. A value which lies in between these limits is in a twilight zone, for which no clear-cut decisions may be made without further analysis.

6.6.4 Critical regions

To illustrate these ideas, consider Figure 6.9 which displays the sampling distribution of some unspecified test statistic if the null hypothesis is correct. For simplicity, assume that this distribution is Normal. The areas under the curve correspond to 95 per cent and 99 per cent confidence intervals. Two types of area should be noted:

1 The central areas which lie within the confidence limits of the test.
2 The areas between the confidence limits and the tails of the distribution. These are critical regions and are concerned with the rejection of the null hypothesis.

As the confidence interval widens, the sizes of the critical regions contract. At a 5 per cent significance level, the area contained between the confidence limits corresponds to 95 per cent of the values of the test statistic. Five per cent of the values lie outside this area in the critical regions. If the test is made more rigorous by testing at the 1 per cent level, the area contained between the limits expands to leave only 1 per cent of the values of the test statistic in the critical regions. In interpreting any test, researchers should remember that extreme values for a statistic can occur by chance. Thus at the 1 per cent level, the chances of rejecting the null hypothesis when it is in fact correct will be once in every hundred tests. At the 5 per cent level, this chance will have risen to five in every hundred.

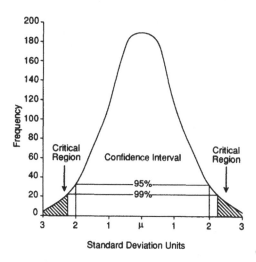

Figure 6.9 Confidence intervals in a Normal test statistic

The critical regions in Figure 6.9 have been drawn at both ends of the distribution. For a 95 per cent confidence interval, this means that 2.5 per cent of the observations will be located in the critical regions at either end. However, if direction is specified in an alternative hypothesis, the critical regions will alter, even though the rigour of the test may be unchanged. Figure 6.10 displays the three critical regions associated with the three alternative hypotheses listed previously.

Tests of null hypotheses may be developed for the following types of problem:

1 Tests concerned with a single mean value.
2 Tests concerned with two or more means.
3 Tests concerned with proportions.
4 Tests concerned with matched pairs (that is, observations made on the same subject at two different periods, for example before and after training).

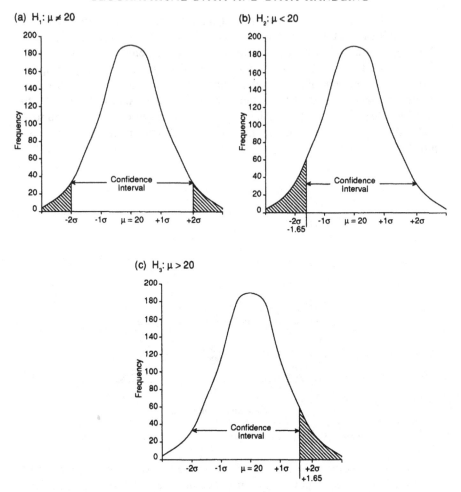

Figure 6.10 Critical regions

All four of these types of problem may be tested using formulae derived from the Normal curve or the *t* test. However, these tests explicitly assume that the population is a Normal shape, at least in large samples. If such an assumption is untenable, a series of non-parametric or distribution-free tests may be applied instead (the latter term is misleading because distributions are involved, but are not stated explicitly). Further details may be found in, among others, Blalock (1979) and Chatfield (1983).

6.7 LIKELIHOOD INFERENCE

The estimation and inference procedures described so far are based on the so-called 'frequentist' concept of probability (Chapter 5, section 5.10).

This postulates that in a repeated series of experiments the probability of an event occurring will tend towards a common value. In practical terms, this means that if a series of different data sets were obtained from the same source, the probability associated with some specified characteristic would be found to be relatively constant. Unfortunately, it is only rarely that researchers obtain more than one data set, and so, only rarely, that the frequentist assumptions can be demonstrated empirically (see Hacking 1965, 1975 for further details). Ideally, what is required is a mode of inference which allows acceptable estimators to be generated using the information contained in the single sample. This implies that the only information which is needed is already contained in the observed data. Such a possibility is provided by likelihood inference.

The principal assumption underlying likelihood inference is the belief that different types of population generate different types of sample. Therefore, the information in a given sample automatically identifies its own population from among the alternatives available. By maximising the information content of the sample data, researchers are assumed to be able to describe characteristics of the population and assess hypotheses concerning it. In other words, the same sorts of descriptive and inferential procedures outlined previously may be applied, but based on a radically different conception on the role of the sample.

6.7.1 Likelihood and log-likelihood functions

In order to maximise this information some procedure for identifying it formally needs to be developed. In much the same way that probability information may be organised into a probability function, so it is possible to express sample information as a likelihood function (Sprott 1973; Pickles 1985). This depends on the observed data and on the unknown parameters associated with them. As the data are essentially fixed in value, given by the sampling process, the only components which may vary in order to maximise the information content available are the parameters. In likelihood inference it is thus assumed that the estimators which are most likely to have generated the observed data are those which maximise the likelihood function.

In order to clarify the basic idea, consider the likelihood function for a binomial variable whose two 'outcomes' are r and $n-r$ (similar examples based on this are presented in Sprott 1973, Chatfield 1983 and Pickles 1985). The likelihood function for this process may be written as follows:

$$L(\theta) = p^r (1-p)^{n-r} \tag{6.14}$$

with:

p the probability of observing outcome r

$(1 - p)$ the probability of observing outcome $n-r$

n the total number of binomial trials

r the number of trials where r is the outcome

The equivalent probability function for this variable is:

$$p(\theta) = p^r (1 - p)^{n-r} n! / r! (n - r)! \qquad (6.15)$$

Notice the similarity between the two equations, but also that the latter includes terms representing permutations of the data (see Chapter 5 for details of permutations).

The likelihood function is somewhat simpler than its equivalent probability function, but further simplicity is provided, if the log-likelihood function is used instead:

$$\log L(\theta) = r \log p + (n - r) \log (1 - p) \qquad (6.16)$$

For given values of r and n, a variety of different values may be generated for the likelihood and log-likelihood functions merely by altering the value of p. It is assumed that the value of the parameter which is the most likely to have generated the data is that which produces a maximum value for both functions.

The values of both functions generated for a random sample of twenty binary observations in which $r=14$ are given in Table 6.7. This summary information may also be presented graphically, as a graph of the likelihood function (Figure 6.11) and the log-likelihood function (Figure 6.12). Notice that the values of the likelihood and log-likelihood functions are convex, sharply-sided, and peak when the probability value is 0.7. This maximum is equivalent to the proportion of observations in the sample of twenty which are classed as category r (i.e., $r=14$, $p=14/20 =0.7$). This implies that the most likely value for r in the population is identical to that

Table 6.7 Likelihood and log-likelihood values for given values of p

p	$1-p$	Likelihood	Log-likelihood
0.0	1.0	0	$-\infty$
0.1	0.9	0.53E–14	−32.86835
0.2	0.8	0.43E–10	−23.87097
0.3	0.7	0.56E–08	−18.99565
0.4	0.6	0.13E–06	−15.89303
0.5	0.5	0.95E–06	−13.86294
0.6	0.4	0.32E–05	−12.64930
0.7	0.3	0.49E–05	−12.21729
0.8	0.2	0.28E–05	−12.78063
0.9	0.1	0.23E–06	−15.29056
1.0	0.0	0	$-\infty$

Note: Based on 20 observations in which 14 are in category 1 and 6 in category 2

156

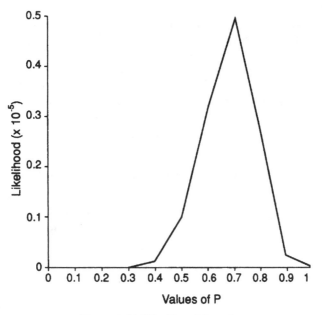

Figure 6.11 Likelihood function

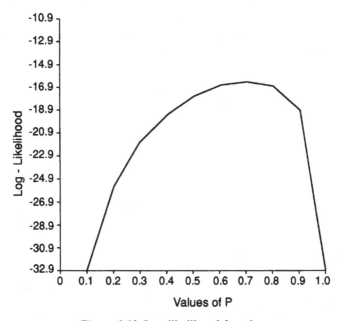

Figure 6.12 Log-likelihood function

observed in the sample. Given limited information, this seems only reasonable.

Both graphs show that many of the probability values (for example, $p=0.2$) could not possibly have generated the data because they lie far from the maximum. However, a number of probability values produce log-likelihood values which are very similar to the maximum. Mathematical calculus needs to be used in order to distinguish between these, and to provide a point estimate for the maximum value of the log-likelihood function. (This is not actually the case for a simple example such as this, but would certainly apply in more complex analyses.)

6.7.2 Relative likelihood and relative log-likelihood

The values of the likelihood and log-likelihood functions at their maxima depend on the data and not just on the parameter estimate. This may be shown by comparing the log-likelihood values in Table 6.7 with those produced by Pickles (1988b) for a smaller sample of hypothetical holiday homes in West Wales (Table 6.8). This data set contained ten binary observations, seven of which were categorised as r. In other words, the only difference between the two sets of data is the size of n. Notice that the values of the log-likelihood produced in the smaller sample are half those of the larger, subject only to rounding errors. However, in spite of these differences, the shapes of the two log-likelihood functions are identical, so it is not possible to assess rival values of p solely from sample size.

The assessment of alternatives near the maximum is made possible using two functions which have standardised maxima. These are the relative

Table 6.8 Comparison of log-likelihood values from Table 6.7 and Table 2 of Pickles (1986)

p	$1-p$	Pickles's data	Table 6.7 data
0.0	1.0	$-\infty$	$-\infty$
0.1	0.9	-16.43	-32.87
0.2	0.8	-11.94	-23.87
0.3	0.7	-9.50	-19.00
0.4	0.6	-7.95	-15.89
0.5	0.5	-6.93	-13.86
0.6	0.4	-6.32	-12.65
0.7	0.3	-6.11	-12.22
0.8	0.2	-6.39	-12.78
0.9	0.1	-7.64	-15.29
1.0	0.0	$-\infty$	$-\infty$

Note: The values displayed in Table 6.7 are twice those in Pickles (1985), given rounding errors, reflecting the size difference between the two samples

likelihood and relative log-likelihood functions. Both are based on the ratios of their respective likelihood and log-likelihood functions at any given parameter value with the maximum values. Thus relative likelihood is defined as:

$$R(\theta) = L(\theta) / L(\theta_{max}) \qquad (6.17)$$

whereas relative log-likelihood is defined as:

$$\log R(\theta) = \log L(\theta) - \log L(\theta_{max}) \qquad (6.18)$$

This latter function is particularly useful because it measures the departures of any particular value of p from the most likely value. Its maximum value is 0, and this is found at the point where p is greatest.

The relative log-likelihood values for the data in Table 6.7 are given in Table 6.9. These are twice the values produced by Pickles (1986) for his smaller sample. However, though these values are related to sample size, the shapes of their distribution functions are different. Figure 6.13 illustrates three relative log-likelihood functions for $r=0.7$, based on samples of ten, twenty and one hundred. As the sample size increases the curves become distinctly more peaked, indicating the inadequacy of values of p thought plausible from the results of the smaller samples.

The value of these relative log-likelihood graphs is that they allow researchers to place confidence intervals around the maximum value. Pickles notes that a horizontal line drawn across the relative log-likelihood functions at -2 distinguishes between the maximum value and those which are only 1/7 as probable. Notice that the values of the function which cross this hypothetical line are much nearer each other in larger samples than in smaller ones. This mirrors the finding given previously for the sampling distribution of the mean.

Any line drawn across from the vertical axis corresponds to a confidence

Table 6.9 Relative log-likelihood values

p	$1-p$	Pickles's data	Table 6.7 data
0.0	1.0	$-\infty$	$-\infty$
0.1	0.9	-10.32	-20.65
0.2	0.8	-5.83	-11.65
0.3	0.7	-3.39	-6.78
0.4	0.6	-1.84	-3.67
0.5	0.5	-0.82	-1.64
0.6	0.4	-0.21	-0.43
0.7	0.3	0.0	0.0
0.8	0.2	-0.28	-0.56
0.9	0.1	-1.53	-3.07
1.0	0.0	$-\infty$	$-\infty$

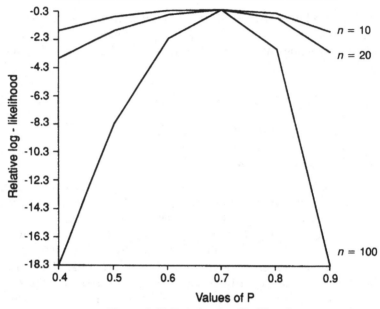

Figure 6.13 Relative log-likelihood

interval. More formally, it corresponds to the $100(1-\alpha)$ per cent interval, where α is a value for which the relative log-likelihood value is equal to

$$\chi^2_{1,\alpha}/2$$

This expression corresponds to an area of rejection identified under the chi-square distribution. For a 95 per cent confidence interval, the expected value under the chi-square distribution for one degree of freedom is 3.84 (see Chapter 5). The value of the relative log-likelihood associated with this is -1.92 and corresponds to the following values of p in the three samples:

	lower limit	*upper limit*
$n=10$	0.39	0.92
$n=20$	0.48	0.84
$n=100$	0.61	0.78

The ability to compare models using the chi-square distribution is particularly important for testing generalised linear models. Dobson (1983) shows that the following expression:

$$-2\,RLL_a = -2\,(LL_a - LL_{max}) \tag{6.19}$$

where

RRL_a is the relative log-likelihood value at point $p=a$
LL_a is the log-likelihood value at point $p=a$, and
LL_{max} is the log-likelihood value at the maximum

160

follows a chi-square distribution for given degrees of freedom. This expression is termed the likelihood ratio statistic. Its use will be described in more detail in Part II. (Barndorff-Nielsen and Cox (1984) show that this statistic is robust under many circumstances but may need adjustment to accommodate censored data.)

6.7.3 Maximum likelihood estimation

In section 6.5.3 it was noted that maximum likelihood estimation may be used to provide acceptable estimators of population parameters. The maximum likelihood estimator is simply the value which maximises the information in the sample. It may be read off directly from the likelihood or log-likelihood functions given previously as the value of p at the peaks of the curves.

The maximum likelihood estimator for the mean is identical to that for the moments method. There is thus no difference in the value of the estimate obtained using either alternative. However, there is a slight difference between the two formulae for the variance in that the maximum likelihood estimator divides through by n rather than $n-1$. This leads to a biased estimator. In spite of this, the estimator may be shown to be consistent in large samples under a variety of conditions allowing robust confidence intervals to be generated. However, Royall (1986) notes that there may be conditions when this does not hold, and an alternative procedure (the delta method) which is consistent in a wider set of circumstances may be chosen instead.

The idea of likelihood estimation may be extended to descriptive and inferential models in exactly the same way as the moments procedure. However, before it may be applied to calculate the values of α and β in the equation for a line (equation 6.7), an additional assumption needs to be made to the two listed earlier. In addition to independent errors and constant error variance, it is necessary to assume that the errors are independent drawings from a Normal distribution (this assumption may also be applied to the y terms instead of the errors). This assumption is needed in order to form a likelihood function for the model.

The probability function for y_i in the equation for the line may be written as:

$$p(y_i) = 1/\sqrt{2\pi\sigma^2} \exp\left[-1/2\sigma^2 (y_i - \alpha - \beta x_i)^2\right] \qquad (6.20)$$

where the values of the probabilities are functions of the data, the two parameters, and the variance of the errors. Using this, it is possible to calculate the probability that any given value of y will occur. However, as this can already be deduced from the data, a more interesting question concerns the behaviour of p given the observations.

To assess this, it is useful to generate the equivalent likelihood function for the model:

161

$$L(y_i) = \hat{\pi} (1/\sqrt{2\pi\sigma^2}) \exp[-1/2\sigma^2 (y_i - \alpha - \beta x_i)^2] \qquad (6.21)$$

where π represents the product of n independent factors.

This expression is a function of the data, the parameters and the variance. It describes the relative odds of obtaining the observed data for all appropriate values of the parameters given that the model is correct. The value of the parameters which is considered 'best' is that which maximises the likelihood function.

Pindyck and Rubinfeld (1976: 52–3) show that this maximisation is obtained by differentiating the likelihood function with respect to α, β and the variance. As with the binomial example given previously, it is easier to work with log-likelihood functions. The following three equations may be produced:

$$\frac{\delta(\log L)}{\delta\alpha} = \frac{1}{\sigma^2} \Sigma (y_i - \alpha - \beta x_i) = 0$$

$$\frac{\delta(\log L)}{\delta\beta} = \frac{1}{\sigma^2} \Sigma [x_i (y_i - \alpha - \beta x_i)] = 0 \qquad (6.22)$$

$$\frac{\delta(\log L)}{\delta\sigma^2} = \frac{-n}{2\sigma^2} + \frac{1}{2\sigma^4} \Sigma (y_i - \alpha - \beta x_i)^2 = 0$$

The solutions to these three estimating equations (which are similar in concept to the normal equation associated with the moments procedure) are

$$\hat{a} = y - bx$$

$$\hat{b} = \frac{\Sigma(x - x)(y - y)}{\Sigma(x - x)^2} \qquad (6.23)$$

$$\hat{\sigma}^2 = \frac{\Sigma(y - \hat{a} - \hat{b}x)^2}{n}$$

For a and b, these are identical to the least squares estimators produced in section 6.5.3. Notice, however, the biased estimator for the variance.

6.7.4 Quasi-likelihood

The great advantage of the likelihood procedure is that it emphasises the most effective use of the observed sample data in generating estimates. It can be shown that likelihood and log-likelihood functions may be written down for all the theoretical frequency distributions presented in this chapter. From these functions, it is possible to generate statistics which contain all the information needed to produce acceptable estimates, so-called sufficient statistics.

In order to generate a full likelihood function it is necessary to assume a theoretical probability process for the data. The likelihood functions of the Poisson and binomial vary because of the differences between the two distributions. Without knowledge of the full distribution therefore, it is not possible to generate maximum likelihood parameter estimates for either distribution. However, Wedderburn (1974b) and Godambe and Heyde (1987) have shown that for generalised linear models, the estimation of model parameters depends on the mean and variance only, and not on a stated distributional shape. Parameters estimated under these conditions are termed quasi-likelihood estimators.

It can be shown that maximum quasi-likelihood parameter estimates are identical to those produced using least squares. Indeed, Godambe and Heyde (1987: 231) note:

> that the optimality of the least-squares estimate ... depends on assumptions concerning the first two moments of the distribution but is otherwise independent of the distribution or distributional form. Similarly, it can be shown that maximum likelihood estimates and maximum quasi-likelihood estimates are identical.

This does not, however, apply to variance estimators (see McCullagh and Nelder 1983: Chapter 8, for more details).

6.8 SUMMARY

This chapter has introduced a number of apparently complex ideas concerning the handling of survey or field data. The need to generalise results from the confines of one study to the broader situation compels researchers to strive for mechanisms which allow such generalisation. Statistical inference provides a possible mechanism which allows information in specially selected data sets ('samples') to be used to infer the parameters of unseen populations. The estimation of parameters such as the mean and variance provides a useful illustration of the inferential procedure.

However, the ability to describe and infer relationships in populations using statistical models is potentially of even greater value. The types of models which may be written down to describe these relationships depend on the nature of the data, the variables and the relationships being considered. As might be expected, a large variety of models exists. In order to present these, and emphasise the points made here, some form of integrated treatment of statistical modelling is needed. The family of generalised linear models offers a suitable framework with which to proceed.

Part II

GENERALISED LINEAR MODELLING

7

TOWARDS INTEGRATION – GENERALISED LINEAR MODELS

7.1. INTRODUCTION

In Part I, it was shown how a variety of numerical measures and descriptive models could be used to summarise relationships in geographical data sets. The potential value of inferential procedures was also stressed, in that data from a single survey may be used to illuminate, or expand on, findings relevant to an unseen population which have been established in other contexts. The advantage of this is that survey data may be generalised or used to expand knowledge beyond the limited confines of the specific survey.

The techniques described in Chapter 6 are appropriate when the object is to infer the characteristics of population parameters from sample statistics. A series of relatively simple devices offer help here, and, moreover, may be developed into a system for assessing hypotheses about relationships in the population. However, many of these relationships may only be understood effectively in their multivariate context, that is, when attention is paid to the effects of two or more variables operating simultaneously. The simple procedures of Chapter 6 provide a key towards the description of multivariate relationships in populations in that they may be developed to estimate the parameters of statistical models of these relationships. Thus, inferential models of population relationships may be developed in much the same way as it is possible to describe patterns in data using descriptive models.

There are many different types of inferential model in geographical use. Some of these are imports from economics, others owe their origins to developments in psychology, political science, public health and statistics. Coming to terms with this variety of models and their range of applications can be quite daunting. This chapter introduces a neat way of getting to grips with the key ideas which are common to most types of model. Indeed, these may be divorced from their historical roots and re-expressed in a common way using the family of generalised linear models. The purpose of this chapter is to introduce the basic format of generalised linear models and the GLIM computer package which is widely available for fitting them.

Subsequent chapters introduce specific generalised linear models which are of use for particular types of analytical problem.

7.2 TYPES OF INFERENTIAL PROBLEM

The inferential models which may be written down vary in composition and character depending on the types of data involved and the sorts of hypotheses being posed. Essentially, three types of modelling problem may be envisaged, involving the assessment of relationships between:

1 continuous measurements;
2 categorical measurements;
3 mixtures of continuous and categorical measurements.

Given the variety of scales associated with these two classes of measurement (see Chapter 3), it is possible to expand these into fifteen distinct types of numerical problem involving the following combinations of measurements:

1 all nominal
2 all ordinal
3 all interval
4 all ratio
5 nominal and ordinal
6 nominal and interval
7 nominal and ratio
8 ordinal and interval
9 ordinal and ratio
10 interval and ratio
11 nominal, ordinal and interval
12 nominal, ordinal and ratio
13 nominal, interval and ratio
14 ordinal, interval and ratio
15 all four measurement types.

Some of these represent distinct types of modelling problem, for which specific solutions have to be sought. Table 7.1 illustrates some of the possibilities. However, some of the procedures listed, particularly those relevant to categorical measures, are rather dated.

Since the late 1950s, new model-based techniques have been available for the analysis of categorical data. Many of the difficulties of analysing categorical data, either in combination with continuous measurements or on their own, have been resolved or greatly simplified. As a result, the fifteen distinct types of problem may be reduced to four areas of interest:

1 classic linear regression analysis;
2 dummy variable regression analysis;

Table 7.1 A variety of analytical modelling procedures

Variable 1	Variable 2	Appropriate measures/tests
Binary	Binary	Difference of proportions
		Chi-square
		Fisher's exact test
		Yule's Q
		Kendall's tau
Multi-way	Binary or	Chi-square
	Multi-way	Cramer's V
		Pearson's contingency coefficient
		Tschuprow's T
		Kendall's tau
Ordinal	Binary	Wilcoxon/Mann–Whitney test
		Wald–Wolfowitz runs test
		Kolmogorov–Smirnov test
		Wilcoxon matched pairs test
Ordinal	Multi-way	ranked ANOVA
Ordinal	Ordinal	rank order correlation
		Kendall's tau
		Somer's D
		Wilson's E
Continuous	Binary	Difference of means test
Continuous	Multi-way	ANOVA
		Intra-class correlation
		Kruskal–Wallis 1-way ANOVA
		Friedman 2-way ANOVA
Continuous	Continuous	Pearson's product moment correlation
		Linear regression

Source: Based on Table 4.15

3 discrete (categorical) regression analysis;
4 tabular data analysis.

A variety of distinct types of inferential model may be written down to represent these problems. These differ in a number of ways, for example in notation, in their reliance on different strategies to estimate parameters, and in their theoretical development.

Most of these differences are the result of the separate development of the models in different disciplines, for example discrete regression in psychology, biology and economics; tabular analyses in statistics, political science and sociology; and regression in statistics, astronomy and agricultural science. It is, however, possible to avoid these differences by respecifying the models in a generalised, integrated framework. This approach has been tried on a number of occasions (for example, by among others, Scheffé 1959; Grizzle *et al.* 1969; Haberman 1974a). The approach to be presented here is based on the work of Nelder and Wedderburn (1972), whose family of generalised linear models provides a framework which is at once comprehensive and easy to compute.

There are three reasons for preferring the generalised framework to the separate treatment of the models listed above:

1 all members of the family of generalised linear models may be specified in a common notation;
2 the probability processes used with the generalised family for inference are themselves interrelated. They form part of the exponential family of probability distributions (see Figure 7.1), and are distinguished from each other by the form of their canonical parameters (that is, the parameters which characterise or typify each distribution – see Chapter 6, section 6.7);
3 these canonical parameters are easily estimated from the sufficient statistics (statistics containing all the information needed to estimate the parameters) of the sample data using a single computer algorithm. This means that only one program needs to be used to tackle all four analytical problem areas, and as a helpful spin-off for the researcher, only one command syntax needs to be learned.

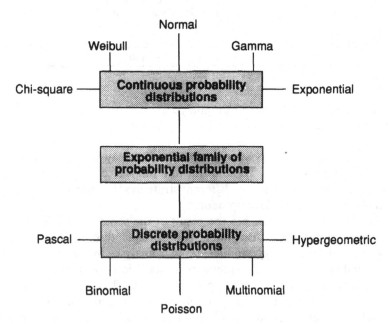

Figure 7.1 Some members of the exponential family of probability distributions

The main effect of these three layers of generalisation is that it is possible to use the generalised family as a practical tool for research and teaching (Aitkin *et al.* 1989). (Appendix A contains further details about the exponential family, canonical parameters and sufficient statistics.)

7.3 COMPONENTS OF A GENERALISED LINEAR MODEL

The family of generalised linear models is popular in statistics and applied research because it is 'extremely simple, and can reflect a justified and widespread desire on the part of the [researchers] to describe the world in the simplest possible manner' (Bibby 1977: 5). All members of the family may be described by a common notation regardless of data type or the purpose of the analysis. At the core of this notation is a simple relationship for an independent random variable, Y. The sample value of this, y, may be decomposed into two parts, one which is predictable, and the other, which is random:

$$y = \text{predictable component} + \text{error component} \qquad (7.1)$$

These parts are related to each other by a mathematical function which is both linear and additive. Algebraically, this may be written as:

$$y = \mu + \varepsilon \qquad (7.2)$$

where

μ represents the part of y whose value is predictable, and
ε represents the part which is a random error

For a set of n random observations (that is, $y_1, y_2, y_3,..., y_n$), equation 7.2 is modified to identify the structure of each:

$$y_1 = \mu_1 + \varepsilon_1$$
$$y_2 = \mu_2 + \varepsilon_2$$
$$y_3 = \mu_3 + \varepsilon_3$$
$$y_n = \mu_n + \varepsilon_n$$

As there is a clear pattern emerging here, some simplification is possible by representing each of the observations in the sample by a subscript, i, which runs in value from 1 to n. Thus the general equation for a sample of n random observations is:

$$y_i = \mu_i + \varepsilon_i \qquad (7.3)$$

In order to turn this general expression into an applicable model, it is necessary to make a series of assumptions about the data and the nature of the analysis to be performed. These involve:

171

1 specifying y from among the many variables collected in the sample;
2 organising the observed data into a form capable of predicting the value of μ;
3 choosing a probability process to enable an assessment of hypotheses and model accuracy.

For the first of these, the choice of y depends on the type of analysis to be performed. For questions 2 and 3, the researchers need to specify a linear predictor, a link function and an error distribution.

7.3.1 Linear predictor

The linear predictor is the term used by Nelder and Wedderburn to refer to the organisation of the observed data in a generalised linear model. For a single observation, it may be presented algebraically as:

$$\eta_i = \beta_0 + \beta_1 x_{i1} + \beta_2 x_{i2} \ldots \beta_k x_{ik} \tag{7.4}$$

where

β refers to k (usually) unknown parameters which have to be estimated, and

x refers to k items of observed data on a single observation (e.g., observation 1) which will be used to predict μ

As an equation of this sort may be written down for each of the n observations in the data set, two simplifying expressions are usually presented instead:

$$\eta_i = \beta_0 + \beta_1 x_{i1} + \beta_2 x_{i2} \ldots \beta_k x_{ik}$$

where each observation is referenced by the $i=1,n$ subscripts, or:

$$\eta = X\beta \tag{7.5}$$

Equation 7.5 uses matrix algebra to simplify the structure of the linear predictor. In this:

η is a $1 \times n$ column vector,
X is an $n \times k$ design matrix representing the linear structure of the observed data for each observation,
β is a $k \times 1$ column vector of parameters requiring estimation.

If the components of matrix X are continuous measurements, their associated parameters scale their effect on β, a simple example being the six-times table given previously, where $\beta=6$ magnifies the effect of a unit change in X six times in η. If, however, the components of X are categorical, their associated parameters represent the effect of each of their categories on η. (Matrix algebra simplifies the presentation of models, but is not in itself an essential component in the description of the generalised family. As a result, it will not be considered in detail here. A review of matrix techniques may be found in, among others, King 1969, Mather 1976, and Hanuschek and Jackson 1977.)

7.3.2 Link function

The linear predictor is related to y by a link function, g:

$$\eta_i = g_i(\mu_i) \tag{7.6}$$

which associates the influences of the observed data with the predictable component of y. This relationship may also be re-expressed using an inverse link function, g^{-1}:

$$\mu_i = g_i^{-1}(\eta_i)$$

As the link (and inverse link) is normally identical for each of the n observations, there is usually no need to distinguish individual links using subscripts.

The link function and linear predictor may be combined to form a basic, or core, equation for the family of generalised linear models:

$$y_i = g^{-1}(\eta_i) + \varepsilon_i \tag{7.7}$$

which is a simple modification of equation 7.2. This may also be written as:

$$y_i \simeq F(g^{-1}(\beta_0 + \beta_1 x_{i1} + \beta_2 x_{i2} \ldots \beta_k x_{ik})) \tag{7.8}$$

or

$$y \simeq F(X\beta)$$

where

> F refers to a member of the exponential family of probability distributions (Figure 7.1), and
> $\simeq$ indicates that the relationship between y and the linear predictor is an approximation

Notice that in equation 7.8 the error component has been omitted. This implies that the assumptions affecting the behaviour of each of the models in the generalised family apply to y rather than the ε component (Collett 1979).

The link functions which are most appropriate for geographical modelling can take a variety of forms. The models associated with the first area of analytical interest – classic regression – take what is termed an identity link. In such a link, $g=1$ so that:

$$\eta_i = \mu_i \tag{7.9}$$

Categorical models, on the other hand, are specified using different links, for example logit and logarithmic link functions. The logit link, which is used with discrete regression models, is expressed as:

$$\eta_i = \log \left(\frac{\mu_i}{\eta_i - \mu_i} \right) \tag{7.10}$$

and the logarithmic link, which is used with tabular analyses involving categorical data, as:

$$\eta_i = \log \mu_i \tag{7.11}$$

Table 7.2 contains a list of some of the most widely-used links associated with generalised linear models.

Table 7.2 Some common link functions in generalised linear models

Link	GLIM notation ($LINK=)	Link function ($\eta=$)
Identity	I	μ
Logarithmic	L	$\log \mu$
Logit	G	$\log (\mu/n-\mu)$
Probit	P	$\phi^{-1} (\mu/n)$
Square root	S	$\sqrt{\mu}$
Exponent	E	$\mu^{**}\{number\}$
Reciprocal	R	$1/\mu$

Notes: The following are default configurations set automatically by GLIM if $LINK is omitted:

Error	Implied $LINK
Normal	Identity
Poisson	Logarithmic
Binomial	Logit
Gamma	Reciprocal

7.3.3 Error structure

The link function and the linear predictor provide the mechanisms for relating the predictable part of y to the observed data. However, these components are not sufficient to turn the generalised family of models into a practical modelling system. A third component is needed which allows the estimated values of y, $\hat{y}$, generated from these models to be compared against theoretical norms or values arising by chance. (This is exactly the same process as applied in Chapters 5 and 6 on assessing hypotheses and the fit of a model.)

The component of generalised linear models which allows this comparison to be made is termed the error structure. In continuous data analysis, the observed predictions from the model may usually be compared for significance against expected values generated by the Normal distribution. For categorical data, the Poisson distribution provides a suitable starting point and is particularly useful for analysing tabular data. Generalisations of the Poisson lead to the binomial and multinomial probability distributions which are suited for analysing proportions. (These characteristics were briefly described in Chapter 5, and are considered again in more detail in Appendix A.)

Table 7.3 lists some examples of generalised linear models derived from equation 7.8 along with their appropriate error distributions and link functions. The seven models presented here are frequently used to analyse geographical problems, as will be shown in subsequent chapters. From this table it can be seen that different types of model may have identical links and errors (e.g., linear regression and the analysis of variance). This feature is fully explained by the components used in their respective linear predictors. For example, models which only include tabulated or dummy observed variables (variables which represent discrete states such as presence or absence of some phenomenon) in the linear predictor are analysed as so-called 'factorial designs' (e.g., analysis of variance and the log-linear model). In contrast, models which are not entirely composed of dummy variables can utilise any available quantitative information directly (e.g., linear regression, logit and probit regression) and there is no need to re-express the model in factorial terms. It is the presence of these tabular variables (frequently termed 'factors' in the statistics literature) which differentiates the two model types. For details of the geographical use of generalised linear models see, among others, Bowlby and Silk (1982), Fingleton (1981, 1983, 1984), Flowerdew and Aitkin (1982), O'Brien (1983) and O'Brien and Wrigley (1984).

Table 7.3 Examples of generalised linear models

Model	Link function	Error distribution
Linear regression	Identity	Normal
ANOVA	Identity	Normal
ANOVA (random effects)	Identity	Gamma
Log-linear model:		
symmetric	Poisson	Logarithmic
asymmetric	Binomial or Multinomial	Logit
Logit regression	Binomial or Multinomial	Logit
Probit regression	Binomial or Multinomial	Probit

Source: After O'Brien (1983) and O'Brien and Wrigley (1984)

7.4 GLIM

7.4.1 Background

The computer package GLIM (Generalised Linear Interactive Modelling) is widely available in British and US universities for fitting generalised linear models to statistical data. It has been developed over many years by the Working Party on Statistical Computing of the Royal Statistical Society and marketed by NAG (Numerical Algorithms Group Ltd, Oxford). It was first released in 1972 primarily for an audience in applied statistics, biostatistics and the natural sciences. Since then it has undergone significant modification and extension in order to widen its appeal, most recently to include a social science audience. The third release – GLIM 3 (Baker and Nelder 1978) – has been installed at over 1,000 university, research institute and commercial sites around the world, and on a wide range of computing systems.

During 1985/6 NAG released two modified versions of GLIM 3 – GLIM 3A and GLIM 3.77 (O'Brien 1986, 1987b). Both of these include enhancements to the basic third release, to simplify its command language, and make it suitable for use with independent microcomputers. An additional feature of GLIM 3.77 is the incorporation of data-handling and manipulation facilities. These allow users to create data tables within the package from external data files and produce a range of summary statistics and visual plots to describe the data prior to modelling. An extra routine allows GLIM 3.77 facilities to be called from outside the package using FORTRAN 77 subroutines supplied by the user.

These enhancements were designed to widen the appeal of GLIM among users of statistical methods, particularly in social science, and to keep GLIM up to date with developments in computing styles and practices. In particular, the advent of powerful microcomputers has allowed a distribution of computing power away from centralised computing departments and towards the users. (For further details on GLIM, see Baker and Nelder 1978, Payne 1986, Reese and Richardson 1984 and Baker *et al.* 1986.)

7.4.2 Specifying generalised linear models in GLIM 3

The core equation of the Nelder and Wedderburn family of models was presented in equation 7.8. The specification of individual generalised linear models, such as those illustrated in Table 7.3, requires modifying this core equation. This is essentially a four-stage design problem involving:

1 defining which of the variables in the data set is to be treated as the response variable (i.e., defining y);

176

2 choosing a suitable probability distribution for y from among the members of the exponential family;

3 specifying the form of the X matrix of independent (explanatory or controlling) variables;

4 choosing a suitable link function which allows the linear predictor to be associated with the predictable component of y.

A summary of the basic steps involved is presented in Table 7.4.

Table 7.4 Summary of specification and fitting commands used in GLIM

Command	Abbreviation	Interpretation
$YVARIABLE	$YVAR	Declaration of the response variable
$ERROR	$ERR	Declaration of the error (probability) component
$FACTOR	$FAC	Declaration of a categorical variable
$CALCULATE	$CALC	Allows internal calculations to be made, including the generation of factor levels
$LINK		Declaration of the link function relating the linear predictor to μ
$FIT		Command to fit a specified model to data
$DISPLAY	$D	Command to display additional information about the fit of the current model

Notes: A full list of GLIM commands is presented in Appendix C.

The syntax needed to set these stages is relatively simple. Stage 1 is set using the $YVARIABLE ($YVAR for short) command, e.g., $YVAR PPT causes the variable PPT to be treated as the response variable within the package. Stage 2 is set using the $ERROR ($ERR) command. The term error is used to refer to the probability distribution which is to be used by GLIM to assess the performance of the fitted model. $ERR P tells GLIM to use a Poisson error process (as might be used when fitting log-linear models).

Stage 3 is rather more complicated. The structure of the explanatory variables can usually be in one of three states:

1 fully continuous measurements;
2 fully categorical measurements;
3 mixtures of both types of measurement.

It is therefore necessary to ensure that each type of measurement is correctly specified within the program. To simplify matters, GLIM has been programmed to assume all variables are continuous unless specified otherwise. If, however, categorical variables are to be used, GLIM needs to be told their names, how many levels each has, and which observations in the raw data are associated with each level. This information is set using

$FACTOR ($FAC) and $CALCULATE ($CALC) commands: for example, $FAC SEX 2 creates a binary categorical variable SEX, whereas $FAC OCC 4 creates an occupation variable with four levels. Note the use of the term factors here to specify categorical measurements. (Readers should not confuse this term with that used in multivariate procedures such as factor analysis, which are not considered in this book.)

Finally, stage 4 is set using the $LINK command, for example, $LINK LOG is used to create the correct logarithmic link structure for fitting log-linear models within the package. (The symbol $ is a system prompt which is used to start and finish command lines in GLIM. The exact form of this prompt may vary on different computer installations.)

7.4.3 Fitting models to observed data

Once the model structure has been completed, the $FIT command may be used to fit the model to the observed data. GLIM produces maximum likelihood estimates for the parameters of this model using a technique known as iterative weighted least squares. The logic of this approach is summarised in the next section.

In response to the $FIT command, GLIM generates the following items as output:

1 cycle information
2 (scaled) deviance
3 degrees of freedom
4 scale parameter

Figures associated with these provide four items of information relevant to the fit of the model. Cycle refers to the number of iterations or successive approximations which were performed by GLIM to produce the maximum likelihood parameter estimates for the data. It works by producing an initial guess at the values of the estimates and then successively revising them until no major changes are detected. (Scaled) deviance indicates the degree of fit of the model using a computed measure whose exact interpretation depends on the model in question. (The difference between scaled deviance and deviance is discussed in section 7.4.5.) Degrees of freedom lists the number of independent items of information associated with the fitted model. The scale parameter is a term which is used in the calculation of the residual variance (the measure summarising the 'unexplained' part of the data). It may either be calculated from the data, or set to an arbitrary value internally, depending on which error distribution has been selected. Taken together, the (scaled) deviance and degrees of freedom provide the information needed to assess the observed fit of the model with the fit which could be expected by chance. Should GLIM be unable to obtain maximum likelihood estimates for some reason, a message to that effect is printed instead.

Further descriptive details associated with the fitted model may be displayed using the $DISPLAY command. A series of options may be selected using this which show, for example, the fitted values (the estimated values of each μ component given the model), the observed values, and the structure of the linear predictor. A fuller list of these options is given in Table 7.5.

Table 7.5 $DISPLAY command options

Option letter	Displays
E	parameter estimates, standard errors, parameter names
R	data, fitted values, standardised residuals
V	covariance of estimates
C	correlation of estimates
S	standard errors of difference of estimates
W	as R, but restricted output
L	the linear predictor
M	current model
A	as E, but extended output
U	as E, but restricted output
D	scaled deviance or deviance and degrees of freedom
T	details of the working matrix

Note: The syntax of this command is: $display options

7.4.4 Estimating model parameters

Maximum likelihood estimates are obtained for the parameters of generalised linear models fitted in GLIM using an iterative weighted least squares algorithm. This procedure is based on a variant of probit analysis, and also on the procedures used in log-linear modelling (Plackett 1974: 72). It is described in detail in Nelder and Wedderburn (1972), and short descriptive summaries are given in Baker and Nelder (1978, Part 1: 1–5), and Baker *et al.* (1986, Appendix A).

It is not appropriate to describe the procedure in any detail here. However, the following points should be noted by the non-technical, non-specialist user of GLIM (a schematic summary is also presented in Figure 7.2):

1 The procedure used for estimating parameters in GLIM generally requires that the raw data observations be used as first approximations to μ, which is used to solve for $\hat{\beta}$, the estimates of the model parameters.
2 In the majority of cases significant improvement may be made on these first approximations by recalculation. This is required because $\hat{\mu}$ and $\hat{\beta}$

are actually functions of each other. It is therefore not possible to obtain the most appropriate estimates for either of these simultaneously.

3 To overcome this, an iterative sequence of successive approximations is applied until the differences in estimates between steps are too small to be significant. At this point, the algorithm used in GLIM is said to have converged and the estimates are presented.

4 These estimates can be shown to be maximum likelihood estimates. Convergence is usually fairly rapid. As a result, GLIM has been programmed to allow ten iterations as default. When this number is reached, a message to the effect that convergence still has not been achieved is printed. Users can overrule this default using the CYCLE and RECYCLE commands (Appendix C.)

5 On occasion, convergence will not occur. Indeed, divergence is potentially possible. Some of the reasons for this are presented in section 7.4.5.

The procedures which are applied to estimate the parameters of fully continuous measurements are slightly different from those applied to categorical measurements. The reason for this is that most categorical data

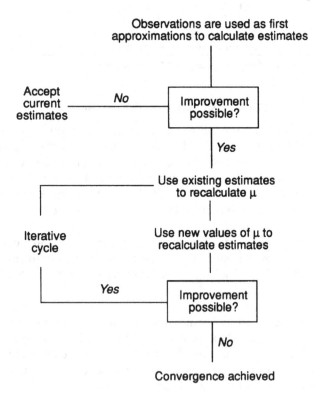

Figure 7.2 A schematic description of the GLIM estimation algorithm

180

matrices contain insufficient independent information to estimate a full set of parameters (see Payne 1977 for details). A matrix in this state is said to be of less than full rank. To estimate parameters for this sort of problem a series of constraints needs to be imposed. Unfortunately, the estimates which are obtained for these constrained parameters depend on the constraints applied (Holt 1979). The system applied in GLIM is termed aliasing. It is described by example in Chapters 8 and 9, and in theory in Appendix A.

7.4.5 Assessing fit

The goodness-of-fit of a generalised linear model requires the assessment of two distinct questions:

1 model adequacy: is the model a suitable description of the population relationship?
2 model accuracy: given (1), is the model as accurate as it could be in terms of reducing uncertainty?

The answer to the first part may depend on prior knowledge, a preliminary analysis of the data to search for relationships and simple descriptive models, or a combination of both. For example, it is worthwhile considering the appropriateness of a particular probability process to represent the surveyed data as the characteristics assumed by the distribution in question may only partially approximate them. Some evidence for this can be gathered from the raw data themselves; for example, if they range over both positive and negative values, then they would not be correctly approximated by the gamma distribution, which is constrained to positive numbers. Similarly, the use of the Poisson to approximate continuous measurements would be inappropriate if the mean and variance measures were not identical or proportional to each other.

The moral is that although a variety of alternative probability processes may be applied to observed data, only those for which an effective justification is forthcoming should be used. This applies in particular if a probability process is used outside the confines of its traditional role, as in the Poisson example. Such an action may not be appropriate; it depends on the characteristics of the data (see Chapter 10 on Poisson regression for an example).

A preliminary analysis of the data is also valuable in assessing the general adequacy of the model to be fitted. Such an analysis is capable of identifying features of the observed data which are odd or unusual in some way, for example, outliers. Some of the procedures relevant to a preliminary analysis were described in Chapter 4. However, in applying generalised linear models to data, two further features need consideration:

1 Does the fitted model represent the mean–variance relationship correctly?
2 Does the fitted model produce additive effects on the scale chosen for the link function?

Both of these reflect the fact that estimates generated by GLIM depend primarily on mean and variance considerations rather than on a complete description of their distributions. If the mean–variance relationship is not modelled correctly (for example, if the data are not independent) then the model is essentially misspecified. The latter reflects the fact that the relationship being hypothesised is linear in a particular scale. This may be assessed graphically. If this shows a non-linear relationship, or some other peculiarity, then the model needs to be reconsidered. Suitable graphical displays for both of these issues are illustrated in Chapters 8 and 9, and described in more detail in Cox and Snell (1968), Belsley *et al.* (1980), Pregibon (1980), Nelder and Pregibon (1983), and Jones (1984).

A further consideration expands on the concept of robust estimation, as illustrated in Chapter 4. It is possible that the estimates provided from the model once it has been fitted to the data are unduly influenced by a subset of the observations, including a single observation. Graphical techniques generally do not highlight this problem, but a number of alternative procedures may be able to detect it. For further information on this, see Hoaglin and Welsch (1978), Atkinson (1981) and Cook and Weisberg (1982).

Assuming the adequacy of the model, the second aspect of goodness-of-fit concerns the accuracy of the components included in it. In general, the problem reduces to assessing which of the variables referenced in the design matrix should be included, and which excluded. This is a rather complex issue because as Seber (1977: 350) points out:

> since there are two possibilities ... 'in' or 'out' of the equation, there are 2^k such (equations). For large k (eg, $2^{10} = 1024$) we are faced with comparing a large number of equations so that we need, first, an efficient algorithm for generating all the possibilities and, second, a readily computed measure for comparing the predictive usefulness of the different models.

Seber suggests the following ways of generating all possible models: sweeping techniques, Hamiltonian walks (Efroymson 1960; Garside 1965), hierarchical trees (Furnival 1971; Furnival and Wilson 1974), and Householder and Givens transformations. Within GLIM, the nature of some of the models which may be fitted suggests the value of the hierarchical approach.

The problem of accuracy essentially involves assessing which of the two models which could represent the data, actually should. To assist with this, Baker and Nelder (1978) suggest that researchers consider the following five types of model:

1 A full model – a model containing as many linear independent parameters as observations and in which the maximum likelihood estimates generated are identical to the raw observations.

2 A null model – a model in which the maximum likelihood estimates are equal for all n observations.

3 A minimal model – a model containing parameters which theory or common sense suggests ought to be included.

4 A maximal model – a model containing all but those terms theory or common sense suggests are inappropriate.

5 The current model – the model associated with the last $FIT command.

The first two of these are essentially redundant because they fail to simplify the information in the raw data. The full model merely reproduces it exactly as a series of parameters, the null model essentially ignores it. The current model therefore needs to be steered towards finding a parsimonious representation of the data which lies somewhere between the minimal and maximal models. A useful 'rule of thumb' in model assessment is to assess the effect of including (excluding) parameters to the (from the) minimal (maximal) model. If it can be shown that the included (excluded) parameters reduce (increase) uncertainty significantly then the evidence suggests that they should be included in the final model. In order to assess this, a measure of reasonableness is required.

This measure of reasonableness can be developed by considering the effect on the random error component of a generalised linear model with varying numbers of terms in its linear predictor. A model with as many terms as observations, a full model in the context just described, will produce estimated values which are identical to those observed. In terms of the core equation for the generalised family this means that all the variability in the observations for the response is attributed to the systematic predictable component, that is, the influence of the error component is reduced to zero.

This feature typifies the full model. Any other model will attribute only some of the variability in the y terms to the systematic part; some will also be attributed to ε. As a result, the expected values given the model will not be identical to the observed values. Discrepancies will be observable when the observed and expected values for each observation in the sample are compared. As the null model represents a situation in which these discrepancies are maximised, it is clear that there is a direct and opposite relationship between maximising the likelihood and minimising the discrepancies. GLIM uses this relationship to suggest a measure of reasonableness termed deviance.

The following relationship can be shown to provide a measure of fit based on the discrepancies of a model which also has the attractive property of being able to infer its appropriateness:

$$-2 (l_c/l_f) \tag{7.12}$$

where

l_c is the likelihood associated with the current model
l_f is the likelihood associated with the full model

A variant, based on log-likelihoods, is

$$-2 \log (l_c - l_f) \tag{7.13}$$

which can be shown to be distributed either exactly as chi-square, or as an asymmetric approximation, depending on the model being fitted (Dobson 1983). Two advantages may be offered for this measure over a number of alternatives which are available:

1 it can be created directly as a result of maximum likelihood estimation;
2 it can be partitioned into component parts so that different types of linear structure may be compared against a common baseline.

Indeed, comparisons need not be made just with the full model, as a maximal model, or some suitable aggregate, may be used instead. (The advantages provided by this will become more readily apparent in the next chapters.) Other types of measure which could be used instead are:

1 the generalised Pearson chi-square statistic;
2 the Wald statistic;
3 the Lagrange multiplier test.

These are described in more detail in Buse (1982) and Pickles (1986).

The interpretation of this deviance measure, which is generated in response to a $FIT command, depends on the model being fitted. In a regression model (to be described in Chapter 8), it essentially corresponds to the sum of squared residuals from the fitted model. It may be tested for significance using the Normal distribution or the chi-square distribution for which its distribution is exact if the model is true. However, in order to do the latter, attention must be paid to scale the deviance by a term representing the variance of the regression residuals. This summary measure is estimated automatically by GLIM and displayed as the scale parameter.

In comparison, if a Poisson error function has been specified the deviance represents a measure termed G^2 by, among others, Bishop *et al.* (1975). This is not the same as the residual sum of squares associated with regression and so cannot be interpreted in the same way. Deviances generated from such an error model are only asymptotically distributed as chi-square. The scale parameter associated with this type of generalised linear model depends on only one unknown (see Chapter 5 and Appendix A). Consequently it does not need to be estimated as is the case with Normal errors. It is set within GLIM to a default value of 1. Notice that in this situation the deviance and scaled deviance measures will be identical because:

$$\text{scaled deviance} = \frac{\text{deviance}}{\text{scale parameter}} \qquad (7.14)$$

The procedure used within GLIM to estimate the scale parameter when it is not set to its default value is based on the mean deviance of the current model. That is, the scale parameter is equivalent to:

$$\text{scale parameter} = \frac{\text{deviance}}{\text{degrees of freedom}} \qquad (7.15)$$

7.4.6 Summary

The themes presented in the previous subsections may now be brought together in summary:

1 The family of generalised linear models possesses common characteristics which may be developed into an integrated framework for studying a variety of linear models.
2 The GLIM package provides for the specification, estimation and assessment of this variety of models using a single command language.
3 The technique of iterative weighted least squares is used to generate maximum likelihood parameter estimates for any member of the generalised family.
4 An iterative procedure is adopted because a direct solution is usually not feasible.
5 GLIM returns information on the fit of a model in the form of a deviance measure and scale parameter. These can be used to generate a scaled deviance statistic which, depending on the model, is either exactly or approximately distributed as chi-square.
6 The value of the deviance measure is that it can be used to assess the effects of different parameterisations of the model in a straightforward way.
7 Other descriptive information may also be generated, for example, information on parameter estimates, residuals and elements of the working matrix. These may be used to investigate the behaviour of a fitted model in greater detail.

The last point will be taken up by examples in the next two chapters. What is important to remember here is that though considerable simplicity is offered by generalised linear models, assumptions are being made. These tend to vary depending on the exact form of the model. Once again, some of these will be examined in the next two chapters.

7.5 CONCLUSIONS

This chapter has shown how it is possible to specify a range of different linear models, suitable for different types of analytical problem, within the same statistical framework. The generalised framework offers considerable economies of effort in studying linear models, and is made all the more powerful by the popularity and availability of GLIM. In the remaining chapters some of the more important generalised linear models in geographical use are described and fitted using data from published research.

8

GENERALISED LINEAR MODELS FOR CONTINUOUS DATA

8.1 INTRODUCTION

The analysis of continuous (interval or ratio) data is traditionally important in geography. This chapter considers two statistical techniques for continuous data analysis which may be presented in the form of generalised linear models. These are:

1 The linear regression model.
2 The analysis of variance (ANOVA).

Both are widely used to model relationships between a so-called 'response' (dependent or endogeneous) variable and a series of 'explanatory' (independent, exogeneous, or controller) variables. These latter variables may vary both in number and in their measurement, including both continuous and categorical data.

8.2 THE LINEAR REGRESSION MODEL

8.2.1 A simple example

The linear regression model is most valuable for modelling 'dependency' relationships between a continuous response variable and one or more explanatory variables. By dependency, we mean that the patterns observed in the response variable can be described by looking at how they relate to changes in the values of the explanatory variables. These changes may correspond to social or physical processes which influence or condition the behaviour of the response variable. However, it is not assumed that the behaviour of the explanatory variables causes the variability in the response variable.

The basic form of this model has already been presented in Part I in connection with the equation for the line:

$$Y = \alpha + \beta X + \varepsilon \tag{8.1}$$

where

 Y represents the response variable
 X represents a single explanatory variable
 ε represents an error term
 α is a parameter representing the intercept or constant term, and
 β is a parameter representing the slope coefficient

Both α and β parameters are generally unknown, and have to be estimated from the observed data using an acceptable estimation strategy. This equation was introduced in Chapter 5 to illustrate the descriptive model which underlies the data in the six-times table. It was also presented in Chapter 6 to illustrate the ordinary least squares estimation strategy. This latter presentation was concerned with the relationship between a population and a sample, and with using sample-based data to approximate population characteristics.

Equation 8.1 corresponds to the population linear regression model. In other words, the model is concerned with describing relationships in the (usually unseen) population. In most cases it is preferable to re-express equation 8.1 as a sample linear regression model:

$$y_i = a + bx_i + e_i \tag{8.2}$$

where

 y_i is the ith sample observation on the response variable Y
 x_i is the ith sample observation on explanatory variable X
 e_i represents a residual: the difference in value between the observed value of y_i in the data, and the value predicted for it, $\hat{y}_i$, given the model
 a is the sample intercept term, and
 b is the sample slope coefficient

An equation identical in form to equation 8.2 may be written down for each of the n observations in the data set.

As was shown in Chapter 6, the estimation procedure known as ordinary least squares is frequently used to provide point estimates of the values of a and b (usually written as $\hat{a}$ and $\hat{b}$) subject to the criterion that the sum of squared residuals from the model is minimised. In other words, in order to provide acceptable estimators of the population parameters, and so describe the relationship between Y and X in the population, a sample model (equation 8.2) is generated and estimates of the sample statistics (a and b) calculated so that the sum of squared residuals is a minimum. Algebraically, this may be written as:

$$\text{minimise } \Sigma\, e_i^2 = \Sigma(y_i - \hat{y}_i)^2 \tag{8.3}$$

If the model is correct, it can be shown that the ordinary least squares procedure will yield estimators which are the best available from among the set of linear, unbiased estimators (see Silvey 1975 for more details).

This type of model may be specified with ease in programs such as GLIM or MINITAB, though the logic of their estimation strategies differs. To illustrate this, consider the transcript of GLIM output presented in Printout 8.1. This shows the commands needed to specify a linear regression model using the six-times table data. The following commands are required to define the model (options specific to this data set are given in brackets):

1 $UNITS – defines the number of observations on the response variable which will be read into GLIM (12 in this case).
2 $DATA – defines the response and explanatory variables to be used in the analysis (EXP and RESP).
3 $READ – allows data entry from the computer keyboard (an alternative command provides data entry from a secondary data file).
4 $YVARIABLE (or $YVAR) – identifies which of the variables defined in the $DATA command is to be used as the response variable (RESP is selected here).
5 $LINK – defines the form of the linear link between the predictable mean of the response variable and the linear predictor (an identity link is required with regression).
6 $ERROR – defines the probability process for the model (a Normal probability distribution is required).

Once defined, the regression of RESP on EXP may now be fitted. The transcript in Printout 8.1 shows that two FIT commands are issued. First, an estimate of the total variability in the response variable is made by fitting a 'null' or 'grand mean effects' model. This is achieved by the command:

FIT

which generates the following information as output:

1 CYCLE 1 – which indicates that the maximum likelihood estimates of this model have been produced directly, that is, without requiring iteration.
2 DEVIANCE 5,148 – which indicates the total amount of unexplained variability in RESP (that is, in terms of ordinary least squares, the total sum of squares for the model).
3 DF (degrees of freedom) 11 – which indicates the number of independent effects remaining in the data after this model has been fitted.

Further information on this fit can be obtained using the $D (for DISPLAY) command, and its E (for estimates) option. This generates the following items of summary information:

189

Printout 8.1 GLIM 3 commands to fit a linear regression equation to the six-times table data

```
£run *glim
  $UNITS 12$.
  $DATA EXP RESP$
  $READ
  1  6
  2 12
  3 18
  4 24
  5 30
  6 36
  7 42
  8 48
  9 54
  10 60
  11 66
  12 72
  $LOOK EXP RESP$
        1        1.000       6.000
        2        2.000       12.00
        3        3.000       18.00
        4        4.000       24.00
        5        5.000       30.00
        6        6.000       36.00
        7        7.000       42.00
        8        8.000       48.00
        9        9.000       54.00
       10       10.00        60.00
       11       11.00        66.00
       12       12.00        72.00
  $YVAR RESP$
  $ERR N$
  $LINK I$
  $FIT$
   CYCLE   DEVIANCE        DF
     1       5148.          11
     $D E$
                ESTIMATE      S.E.      PARAMETER
     1           39.00       6.245      %GM
           SCALE PARAMETER TAKEN AS      468.0

  $FIT EXP$D E$
   CYCLE   DEVIANCE        DF
     1      0.2261E-26      10

                ESTIMATE         S.E.      PARAMETER
     1       -0.1066E-13   0.9255E-14   %GM
     2           6.000     0.1257E-14   EXP
           SCALE PARAMETER TAKEN AS    0.2261E-27
  $STOP
```

Note: £ and $ are operating system and GLIM command prefixes respectively

1 PARAMETER %GM – which indicates that a single parameter has been fitted to the data. This is the %GM parameter corresponding to the grand mean of RESP.
2 ESTIMATE 39.0 – which indicates the value of %GM estimated by the iterative weighted least squares algorithm used by GLIM.
3 SE (standard error of the estimate) 6.245.

The fourth item of summary information which is generated by GLIM in response to the $D command is the SCALE PARAMETER. This is estimated as 468.0 by dividing the deviance value by the number of degrees of freedom (that is, 5,148/11 = 468). This information represents the variance of the residuals from the model and is of use in assessing the overall value of the fitted model to represent the observed data (see next section).

The second model to be specified, $FIT EXP$, includes the observed information on the explanatory variable, EXP, in the linear predictor. In response to this, deviance has fallen from 5,148 to 0 (actually, to 0.2261E−27 but this is so small as to be effectively 0), and the number of degrees of freedom has fallen from 11 to 10. The second $D command shows that the estimated value of %GM, now interpreted as the intercept (or constant) of the linear regression model, is 0 (standard error 0), whilst the estimate of the slope coefficient is 6 (standard error 0). Thus the sample linear regression model for the six-times table data is:

$$RESP = 0 + 6EXP \qquad (8.4)$$

which accords with our previous knowledge.

The information supplied by the maximum likelihood analysis in GLIM may be compared with that supplied for the identical least squares analysis in MINITAB (Printout 8.2). In this, the data are read into two column variables, c1 and c2. The command,

    regress c2 1 c1

is issued, which tells MINITAB to calculate a linear regression analysis using the data in c2 as the response variable (equivalent to RESP in Printout 8.1), and the single variable in c1 as the explanatory variable (equivalent to EXP in Printout 8.1). The estimated model is identical to that produced by GLIM, in spite of the fact that the estimation strategy (and logic) is different. Notice also that the values listed in Printout 8.2 as COEF and STDEV – 0 and 0 for the constant, and 6 and 0 for c1 – are identical to those listed under the ESTIMATE and SE headings in the second fit in GLIM. Similarly, under the heading 'ANALYSIS OF VARIANCE' in Printout 8.2, we can see that the value listed for the TOTAL (sum of squares) is 5,148 for 11 degrees of freedom, and for the ERROR (sum of squares) is 0 for 10 degrees of freedom. These correspond to the values of deviance and degrees of freedom produced in fits 1

Printout 8.2 MINITAB commands to fit a linear regression equation to six-times table data

```
£run *minitab
-read c1 c2
-data
    1   6
    2  12
    3  18
    4  24
    5  30
    6  36
    7  42
    8  48
    9  54
   10  60
   11  66
   12  72
-end
-regress c2 1 c1
```

The regression equation is
C2 =0.000000 + 6.00 C1

Predictor	Coef	Stdev	t-ratio
Constant	0.00000000	0.00000000	*
C1	6.00000	0.00000	*

s = 0   R-sq = 100.0%   R-sq(adj) = 100.0%

Analysis of Variance

SOURCE	DF	SS	MS
Regression	1	5148.0	5148.0
Error	10	0.0	0.0
Total	11	5148.0	

```
-stop
```

Notes: Commands beginning with a £ are system commands relevant to the Durham and Newcastle computers
Commands beginning with a - are MINITAB commands

and 2 of GLIM respectively, though it should be noted that the maximum likelihood strategy used in GLIM is not based on equation 8.3.

8.2.2 Assessing global goodness-of-fit

The object of linear regression is to find a suitable combination of explanatory variables and parameters which best describes the variability in the response variable. In the statistical literature describing regression, the maximum amount of variability to be modelled is frequently termed the total sum of squares (TSS). Algebraically, this term may be written as:

$$TSS = \Sigma (y_i - \bar{y})^2$$

where y_i and $\bar{y}$ are as before. The TSS may be rewritten to reveal two further sums of squares, one assumed to result from the explanatory power of the model – the regression sum of squares (RSS) – and the second to correspond to residual (unexplained) variability – the error sum of squares (ESS):

$$TSS = RSS + ESS$$

or

$$\Sigma (y_i - \bar{y})^2 = \Sigma (\hat{y}_i - \bar{y})^2 + \Sigma (y_i - \hat{y}_i)^2 \tag{8.5}$$

where

y_i corresponds to the $i=1, \ldots, n$ sample observations on the Y response variable

$\bar{y}$ corresponds to their observed sample mean

$\hat{y}_i$ corresponds to the 'expected' value of y_i given the model (that is, the value of y_i predicted by the model for the given explanatory variables)

Following equation 8.5, the estimates of the model are obtained by minimising the value associated with ESS using procedures such as ordinary least squares (MINITAB). The ratio of RSS to TSS provides a measure of the overall fit of the model. (This ratio may also be expressed as 1 minus the ratio of ESS to TSS.) This ratio is termed the coefficient of determination or r^2. If

$$RSS/TSS = 1 - (ESS/TSS) = 1 \tag{8.6}$$

the model is said to describe the observed response data exactly, as the sum of squared residuals has been achieved at ESS=0. If, however,

$$RSS/TSS = 1 - (ESS/TSS) = 0 \tag{8.7}$$

the model is said to fail completely to describe the observed response data. When this occurs, ESS = TSS and RSS = 0. Values between these two extreme points indicate the relative overall performance of the model. For the analysis of the data in Printout 8.2, the value of r^2 is given as 100% (equivalent to 1). This is supported by the fact that the RSS and TSS values (corresponding to the values of SS associated with the Regression and Total entries in the Analysis of Variance table) are identical, and the value of ESS is 0.

GLIM does not produce a value for r^2 directly, though it can be calculated easily by forming the ratio of the value of deviance produced by the second fit command (which corresponds to the amount of unexplained variability in the data – ESS), to that produced by the first (TSS). This ratio is 0. It follows, therefore, that the change in deviance between the two fit commands (that is, between the null model and the linear regression model) corresponds to RSS. The ratio of this to TSS is 1, indicating that the linear regression model fits the observed data perfectly. This difference may also be used to generate a measure termed pseudo-r^2, which is the difference between the two deviance values divided by the deviance of the null model. For regression this is equivalent to r^2. This characteristic does not apply to other models.

8.2.3 Assessing local goodness-of-fit using residuals

The previous discussion has focused on measures such as deviance and r^2 which provide an assessment of the overall or global fit of a linear regression model. While these may suggest that a model fits the data reasonably well, they tell the user nothing about the fit for all combinations of response and explanatory variables. It is possible that a model which is thought to be reasonably good may in fact be quite poor for certain combinations of observed data. It is therefore necessary to supplement these global measures with some form of local assessment. This may be achieved by looking at the residuals from the model.

Regression residuals are usually defined as the differences between the observed values of the response variable and their predicted or fitted values given explanatory information. Both the observed and fitted values are stored by GLIM in so-called 'system variables' (%YV for the observed data and %FV for the fitted data). These are created automatically by the program and so do not need to be specified in a $DATA command. For the six-times table data the values of %YV and %FV can be seen using a $LOOK command (Table 8.1). This shows that the values are identical and that therefore there are no residuals in this analysis. Once again, this accords with our previous knowledge.

The six-times table data are special in that the values predicted by the linear regression model for the twelve observations on Y are identical to those observed in the survey data. This does not apply to the data in Table 8.2 which lists the unemployment rate (UN) and the percentage of economically active males (EA) within British counties in 1981. The regression of UN against EA yields the following model:

$$UN = 39.5 - 0.483EA \qquad (8.8)$$
$$(4.65) \ (-3.41)$$

The values in brackets are t-ratios. They are calculated by dividing the values of the parameter estimates by their standard errors. As a general rule, values greater than 2 are significant at the 5 per cent significance level

Table 8.1 Comparing the fitted values
with the observed values of the response

| $LOOK %YV %FV$ | | |
OBS	%YV	%FV
1	6.000	6.000
2	12.00	12.00
3	18.00	18.00
4	24.00	24.00
5	30.00	30.00
6	36.00	36.00
7	42.00	42.00
8	48.00	48.00
9	54.00	54.00
10	60.00	60.00
11	66.00	66.00
12	72.00	72.00

for a two-tailed test. The overall fit of this model is poor ($r^2 = 15.4\%$) indicating that the explanatory variable, percentage of economically active males per county, is not a particularly powerful descriptor of county unemployment rates. The residuals associated with this model are set out in Table 8.3 and in Figure 8.1. Before we can use these to interrogate the model listed above, we need to know something about the assumptions of linear regression modelling.

8.2.4 Assumptions

It is important to realise that there is a major conceptual difference between the use of linear regression on the six-times table data and the unemployment data. In the former analysis the relationship between *RESP* and *EXP* is a mathematical constant fixed by the nature of the data. The correspondence between the known relationships (α of 0 and β of 6) and their estimates merely reflects the determinism of the example. However, in the unemployment example, there is no *a priori* knowledge of the values of α and β. Consequently, it is not entirely clear how far one should be prepared to go in accepting the fitted model as adequate. In applying regression (or any statistical model for that matter) to social and environmental data, attention must be paid to assessing the suitability of the model to represent the observed data. In particular, is the model a suitable tool to represent the social and physical properties assumed to underlie the observed data?

Linear regression analysis is most useful when the social or physical situation being examined can be conceived in terms of a dependency relationship between a single response variable and one or more explanatory variables. In general, the response variable should be recorded as

195

Table 8.2 Economic activity and unemployment rates for
males by county (in 1981)

EA	UN	County
65.1	6.6	Bedfordshire
66.0	4.1	Berkshire
64.9	4.4	Buckinghamshire
52.0	7.5	East Sussex
61.0	7.7	Essex
61.9	6.6	Hampshire
64.8	4.4	Hertfordshire
51.3	12.1	Isle of Wight
59.2	8.1	Kent
63.7	4.2	Oxfordshire
61.7	0.0	Surrey
56.3	3.8	West Sussex
62.1	5.7	Cambridgeshire
57.4	9.1	Norfolk
59.0	6.1	Suffolk
63.7	8.3	Greater London
60.5	8.1	Avon
52.8	15.3	Cornwall and Isles of Scilly
54.3	11.0	Devon
53.6	7.8	Dorset
60.7	6.6	Gloucestershire
57.5	7.6	Somerset
61.9	6.6	Wiltshire
62.0	8.8	Hereford and Worcester
60.3	10.7	Shropshire
63.5	9.6	Staffordshire
63.1	8.1	Warwickshire
62.7	12.6	West Midlands
60.6	10.8	Derbyshire
64.1	7.1	Leicestershire
58.7	11.0	Lincolnshire
63.0	6.8	Northamptonshire
61.4	11.1	Nottinghamshire
59.6	13.4	Humberside
59.0	8.5	North Yorkshire
59.8	15.9	South Yorkshire
61.3	10.6	West Yorkshire
61.9	10.5	Cheshire
62.2	12.5	Greater Manchester
59.8	11.6	Lancashire
60.5	18.6	Merseyside
59.6	9.1	Cumbria
61.6	18.2	Cleveland
59.3	15.0	Durham
58.5	13.2	Northumberland
60.2	15.8	Tyne and Wear
55.8	13.5	Clywd
54.7	15.5	Dyfed

196

Table 8.2 Continued

EA	UN	County
58.2	13.7	Gwent
53.0	16.8	Gwynedd
57.1	15.9	Mid Glamorgan
56.9	9.7	Powys
59.6	11.1	South Glamorgan
57.0	14.4	West Glamorgan
60.5	8.7	Borders region
62.2	14.7	Central region
59.5	12.5	Dumfries and Galloway region
61.0	14.6	Fife region
62.3	8.6	Grampian region
60.1	15.2	Highland region
63.7	11.3	Lothian region
61.7	16.8	Strathclyde region
60.8	13.1	Tayside region
57.8	12.6	Orkney Islands
64.4	6.8	Shetland Islands
53.9	20.7	Western Isles

Note: EA = Economically active
UN = Unemployment rate

Table 8.3 Residuals from regression of economic activity and
unemployment rates

County	Observed	Fitted	Residual
1	6.600	8.061	−1.461
2	4.100	7.626	−3.526
3	4.400	8.158	−3.758
4	7.500	14.39	−6.890
5	7.700	10.04	−2.342
6	6.600	9.607	−3.007
7	4.400	8.206	−3.806
8	12.10	14.73	−2.628
9	8.100	10.91	−2.811
10	4.200	8.737	−4.537
11	0.0	9.704	−9.704
12	3.800	12.31	−8.512
13	5.700	9.510	−3.810
14	9.100	11.78	−2.681
15	6.100	11.01	−4.908
16	8.300	8.737	−0.4375
17	8.100	10.28	−2.183
18	15.30	14.00	1.297
19	11.00	13.28	−2.279
20	7.800	13.62	−5.817
21	6.600	10.19	−3.587
22	7.600	11.73	−4.133
23	6.600	9.607	−3.007

Table 8.3 Continued

County	Observed	Fitted	Residual
24	8.800	9.559	−0.7588
25	10.70	10.38	0.3200
26	9.600	8.834	0.7659
27	8.100	9.027	−0.9274
28	12.60	9.221	3.379
29	10.80	10.24	0.5649
30	7.100	8.544	−1.444
31	11.00	11.15	−0.1530
32	6.800	9.076	−2.276
33	11.10	9.849	1.251
34	13.40	10.72	2.682
35	8.500	11.01	−2.508
36	15.90	10.62	5.278
37	10.60	9.897	0.7031
38	10.50	9.607	0.8929
39	12.50	9.462	3.038
40	11.60	10.62	0.9784
41	18.60	10.28	8.317
42	9.100	10.72	−1.618
43	18.20	9.752	8.448
44	15.00	10.86	4.137
45	13.20	11.25	1.950
46	15.80	10.43	5.372
47	13.50	12.55	0.9461
48	15.50	13.09	2.415
49	13.70	11.39	2.305
50	16.80	13.91	2.893
51	15.90	11.93	3.974
52	9.700	12.02	−2.323
53	11.10	10.72	0.3818
54	14.40	11.97	2.426
55	8.700	10.28	−1.583
56	14.70	9.462	5.238
57	12.50	10.77	1.734
58	14.60	10.04	4.558
59	8.600	9.414	−0.8138
60	15.20	10.48	4.723
61	11.30	8.737	2.563
62	16.80	9.704	7.096
63	13.10	10.14	2.962
64	12.60	11.59	1.012
65	6.800	8.399	−1.599
66	20.70	13.47	7.228

either an interval or a ratio measurement (though it is possible, under limited circumstances, to use categorical measurements), and be linked to the explanatory variables by a mechanism which is linear and additive in its parameters. Both examples of regression presented so far possess this

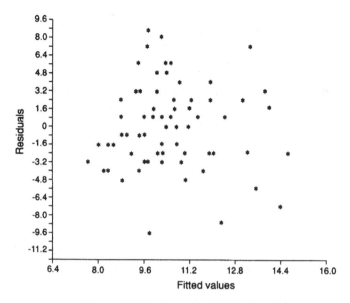

Figure 8.1 Scatterplot of residuals

property. This does not mean that the individual variables need be linearly related, rather that their parameters must be. Thus, models 1 and 2 below could be used as linear regression models, but model 3 could not:

1   $Y = \alpha + \beta X_1 + \beta X_2 \quad\quad + \varepsilon$
2   $Y = \alpha + \beta X_1 + \beta X_2^k \quad\quad + \varepsilon$
3   $Y = \alpha + \beta X_1 + (1/\beta) X_2 + \varepsilon$

In applying regression to social or physical data whose functional form is not pre-determined, as is the case in the six-times table example, it is important to realise that the model is based on assumptions. The number of these assumptions tends to depend on which author one reads, but the following seem to be reasonably common:

1 The model is correctly specified – this is assumed as a matter of course, and implies (a) that variables which should be present are present, (b) that irrelevant variables are not present, and (c) that the error component in the population model is additive.
2 That there is sufficient information in the explanatory variables to allow estimates of their parameters to be made.
3 That the errors possess the following properties: (a) they have an expected value of zero and a constant variance for each of the n observations (an assumption known as homoscedasticity), (b) that none are correlated (an assumption known as no autocorrelation), and (c) that they are Normally distributed.

199

Significant departures from these assumptions tend to corrupt the statistical behaviour of the model and thus compromise its descriptive value. However, though the assumption of Normality is made, it can be shown that the important properties of least squares estimators do not depend on it (McCullagh and Nelder 1983). Of far greater importance are the assumptions of constant variance and lack of correlation. This also applies to quasi-likelihood estimators such as those generated for many of the models in the exponential family. Geographers should note straight away that independence is rarely satisfied by geographical data simply because measurements made in similar locations tend to be more similar than would be expected by chance.

8.2.5 Some graphical checks of assumptions

Graphics provide a useful way of checking that the assumptions of the regression model are not violated to a considerable degree. The most useful graphical techniques use the residuals to check for model inadequacies. Tukey (1977: 125) notes: 'residuals ... are to the data analyst what powerful magnifying glasses, sensitive chemical tests for bloodstains and delicate listening devices are to a story-book detective'. In effect, they provide a most important clue to the appropriateness of the fitted model to describe the patterns and relationships in the observed data.

A regression model which describes observed data adequately should generate a set of residuals which does not display an obvious pattern; in effect, it should be essentially random. By plotting the residuals against the estimated values of the response variable (as was done in Figure 8.1), considerable information about the fit may be obtained. A variety of patterns may emerge in these so-called 'catch-all' plots (Jones 1981, 1984). For example, an adequate model should produce a random pattern (Figure 8.2(a)). The remaining plots in Figure 8.2 illustrate potential assumption violations:

1 An unsatisfactory link function depicted by the non-linear shape of the residuals (Figure 8.2(b)).
2 A violation of the constant error variance assumption by the non-circular shape of the residuals (Figure 8.2(c)). This indicates that the variance of the errors may depend on the magnitude of the observed data.
3 A violation of the no-autocorrelation assumption by the presence of the systematic patterning in the residuals, indicating that values which are similar in magnitude behave rather more like each other than values which are dissimilar (Figure 8.2(d)).
4 A failure to fit the model adequately to all data points, thus ignoring values whose effect on the model may be extreme (Figure 8.2(e)).

(Notice that the assumptions pertaining to the errors in the population model are assessed by examining the residuals from the sample model.)

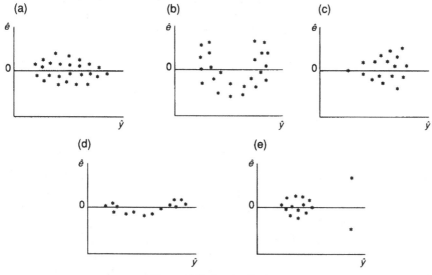

Figure 8.2 Catch-all plots

If these plots suggest that underlying assumptions are being violated it is important to modify the model as its summary measures and measures of fit will not be as good as they could be. For example, if the link appears to be incorrect the source of the violation should be identified and rectified. This may involve transforming the model from its original units of measurement to some alternative, for example natural logarithms, as is frequently done in distance–decay and density–decay models (Taylor 1977). However, as Haworth and Vincent (1979) show, this may create unexpected problems, especially if it alters the functional linkage of the error term (see Kennedy 1979 for a detailed discussion of this problem). Instead of a global transformation of the model, it may be possible to identify particular variables which are involved. This may be especially valuable in the case where there are several explanatory variables, only some of which are responsible for the inadequacy of the link. Jones (1984: 222) notes:

> if . . . a non-linear relationship (exists) between Y and all explanatory variables, then it can be suggested that a transform of Y is required; a curved relationship with only one X suggests that the X variable should be transformed.

The 'ladder of powers' introduced in Chapter 5 provides possible transformations which could be considered. Watts (1981: 80) suggests a variety of other tests and modifications which may be used.

Autocorrelation and heteroscedasticity also cause a regression model to be less than optimal. These problems can be tackled in a number of different ways. For example, if the residuals display a systematic pattern with clusters of positive and negative values in evidence, it may be possible to pinpoint a specific cause and so model it explicitly. In time-series data sets, this cause may reflect yearly, monthly or seasonal influences which are not random and so should not be assigned to the residuals. Alternatively, the data may be divided into subsets each of which are studied individually.

8.2.6 Residual analysis of Table 8.2

Given these assumptions we are now in a position to examine the information contained in the residuals from the regression analyses of Table 8.2. The scatterplot in Figure 8.1 illustrates the covariability of the residuals produced by GLIM (vertical axis) and the fitted values (horizontal axis), that is, the estimated values for the response variable produced by the model. For fitted values between about 8 and 12 the scatter seems reasonably random with no obvious pattern emerging. However, the scatter does seem to fragment somewhat for fitted values greater than 12. Viewed from the bottom right, the whole scatter appears rather like an arrow-head, narrowing towards the bottom right and indicating a potentially inverse relationship between the two variables. This may indicate the presence of heteroscedasticity, but it is important to note that the 'shape' is unduly affected by outlying points which are psychologically more prominent than other points. As a result it is important to check whether these outlying points are peculiar in some way, as their measured values may be unduly important. (An obvious reason why a point could appear peculiar is if it has been recorded incorrectly in the data. A preliminary analysis of the data set should pick this mistake up before analysis.)

One way of assessing whether data points are extreme is to transform the raw residuals of Figure 8.1 and Table 8.3 into standardised residuals. These are defined to have a mean value of 0 and a standard deviation of 1. The advantage of this change in scale is that it provides a common basis for comparing residuals. The information in the standardised residuals is identical to that in the raw residuals, and indeed would form the same pattern. However, because the values of the raw residuals depend on the original units of measurement, large residuals may emerge solely because the original measured values were large. With a common scale, this source of confusion may be removed, as residuals lying outside plus or minus two standard deviation units from the mean can be considered potentially extreme. Four counties fall outside this range: Surrey, West Sussex, Merseyside and Cleveland.

It is possible to generate standardised residuals in GLIM using the following code (after Gilchrist 1983):

$CALC RES=(%YV−%FV)*%SQRT(1/%SC)   (8.9)

where

%YV   is the system vector containing the observed values of the response variable

%FV   is the system vector containing the fitted values given the model

%SC   contains the value of the scale parameter from the fitted model

%SQRT   is a system function to calculate square roots

This expression is a special case of a more general formula which produces standardised residuals for any generalised linear model. This is:

$CALC RES=(%YV−%FV)*%SQRT(%PW/(%SC*VA))   (8.10)

where

%PW   is a vector containing any prior weights used in the fitting of the model

VA   is defined as the variance function. For Normal errors, this is equal to 1, whereas for Poisson errors, it is equal to %FV.

The presence of a spatial structure in these standardised residuals may also be examined by plotting them in a map (Figure 8.3). This shows that there is a distinct spatial pattern to the residuals which directly violates the no-autocorrelation assumption. Positive residuals are distributed throughout most of Scotland, Wales, the North and North-West, Humberside and Cornwall, while negative residuals are found throughout the South and East, East Anglia and the East Midlands. The only counties outside this contiguous area to have negative signs are North Yorkshire, Cumbria, Powys, Borders, Grampians and the Shetlands. As this distinction is quite clear cut, it follows that it ought to be incorporated explicitly within the model rather than be reduced to the residuals. A dummy variable with levels distinguishing between the North and South might prove useful (for further details of how to use dummies, see section 8.3.4).

The initial inspection of the residuals from the regression models seems to suggest that autocorrelation is a problem, and that the data may well be heteroscedastic. A number of outlier observations appear to be present in the data. To overcome their effect, the extreme values might be removed and the analysis repeated on the remaining data values. For autocorrelation, some attempt should be made to model the spatial pattern explicitly, and for heteroscedasticity, some form of transformation or weighted regression analysis might be useful. In addition to these checks, it is also important to check that the link function specified for the model is adequate.

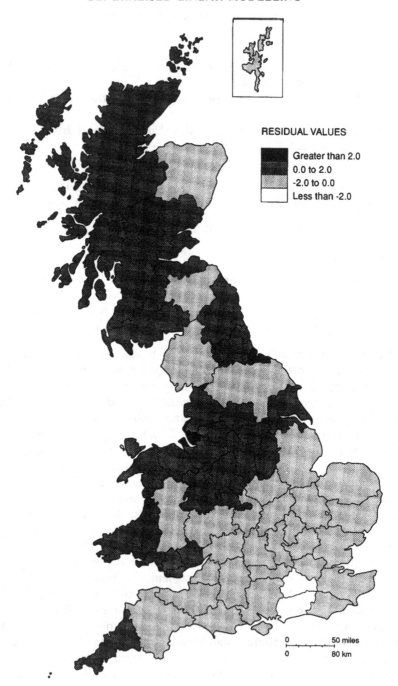

RESIDUAL VALUES

Greater than 2.0
0.0 to 2.0
-2.0 to 0.0
Less than -2.0

Figure 8.3 Map of standardised residuals

In regression analysis, an identity link is specified in which the linear predictor is directly related to the predictable mean of the response variable. If the model is correctly specified, the residuals should be approximately linear when plotted against the order statistics for a Normal distribution. It is possible to check for this using the data in Table 8.2 either by plotting the standardised residuals, or by creating normalised residuals. The latter are produced from the expression:

$$\$CALC\ NR = UN^{**}(1/3) - \%FV^{**}(1/3)/\%FV^{**}(1/3) \tag{8.11}$$

The order statistics are produced from:

$$\$CALC\ OR = \%ND((\%GL(66,1) - 0.5)/66) \tag{8.12}$$

The plot of NR and OR is presented in Figure 8.4. The advantage of this type of plot is that it immediately shows if the relationship is sufficiently linear to continue using regression, and also which of the counties are extreme. This latter feature arises because the residuals are ordered before plotting, positioning the extreme values at either end of the plot. Apart from one observation, the remainder appear to be linearly related, indicating that the functional form is reasonable. The major discrepancy is Surrey. Its normalised residual is particularly extreme and is well away

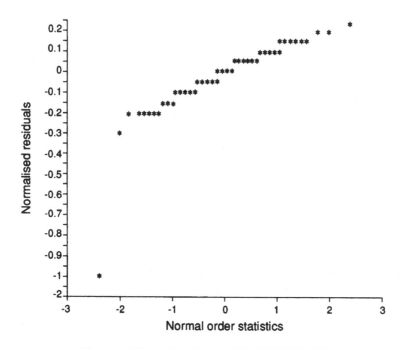

Figure 8.4 Normalised plot of data in Table 8.2

205

from the remainder. From Table 8.2 we see that its value for unemployment is zero. This is significantly different from the unemployment figures for other areas and suggests that the measurement and recording of the Surrey data may be peculiar. In fact, unemployment details for the county of Surrey were suppressed for the period in question, and so it should be excluded from the model.

8.2.7   Influential observations

Residual analysis is useful in identifying observations which are peculiar in some sense given the main body of the data. However, they are less suitable for identifying observations whose influence on the analysis is particularly strong. It is not sufficient to assume that observations which have large residuals are automatically influential. This is because observations which are typical of the data may indeed have large residuals, and the residuals associated with influential observations may be quite small. In order to identify influential observations a new measure is required which can assess the sensitivity of each of the observations in the data set.

One such measure is termed 'leverage'. This is defined to be the diagonal elements in the 'projections matrix': a matrix which compares the covariation in the observed response with its equivalent in the estimated response. This matrix is produced automatically by MINITAB, and the leverages may be stored in a column variable using a sub-option of the regression command:

    regress UN 1 EA, c3 c4;
    subc hi c5.

UN and EA are the column variables containing the response and explanatory variables respectively. c3 and c4 are column variables which are used to store the standardised residuals (c3) and the fitted values (c4). The sub-option specifies that a variable hi (MINITAB terminology for leverages) is stored in c5. By plotting c5 against the row order of the data, an index plot of the leverages is produced (Figure 8.5). The leverages are measured against the vertical. Pregibon (1983) notes that observations whose coefficients exceed $3(p/n)$ are said to be influential (where p is the number of parameters fitted, including the intercept, and n is the number of observations). Thus for the data in Table 8.2, p equals 2 and n equals 66. The critical value for leverage is thus 0.09. There are thus three influential counties – East Sussex, the Isle of Wight and Cornwall/Isles of Scilly – all of which have particularly low values of EA.

In addition to leverage, Pregibon (1982) suggests a coefficient of sensitivity based on a modification of the adjusted residuals from the model. These are calculated from the standardised residuals by dividing each item as follows:

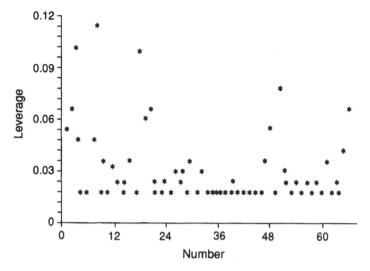

Figure 8.5 Leverage values for Table 8.2

$CALC ADJ=RES/%SQRT(1−H) (8.13)

where

H   is the variance of the linear predictor and is calculated from (%WT*%VL)/%SC. (%WT is a vector of iterative weights, and

%VL is the variance of the link.)

The coefficient of sensitivity (C) is calculated using ADJ and H:

$CALC C=ADJ*ADJ*H/(1−H) (8.14)

These may be plotted for each observation using an index plot (Figure 8.6), in which the coefficients are plotted along the vertical and the county indices are plotted on the horizontal (for further details about this plot, see Belsley *et al.* 1980). These indices are easily generated in GLIM using:

$CALC IND=%GL(%NU,1) (8.15)

where

%NU is the index number for each row of observed data.

8.2.8 Summing up

In this section we have begun to introduce some of the features associated with the geographical use of the linear regression model. The key point to note is that all regression models need to be examined to assess whether

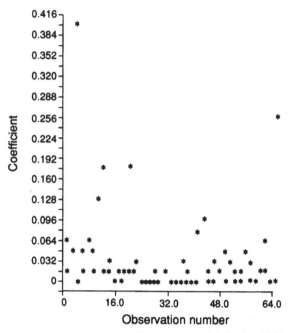

Figure 8.6 Index plot of the coefficient of sensitivity for Table 8.2

they fit the observed data both globally and locally. Global measures are insufficient on their own as they do not describe directly subsets of the data which seem to be peculiar in some way. Residual analysis is crucial in examining the fit of a regression model. Residuals provide a means of assessing whether the assumptions of the model are satisfied by the data or are seriously violated.

As we saw in the determinist example of the six-times table, the values of the response variable predicted by the model are identical to those observed. As a result, there are no residuals to consider. However, geographical data sets rarely display this characteristic. In the unemployment example, we noted how the residuals suggested violations of the assumptions of no autocorrelation and homoscedasticity, and identified the possibilities of extreme data values and influential data values. These need to be considered explicitly if the model is to be improved adequately. Some of these issues are touched upon again in the next section which considers how a regression model might be developed to cater for several explanatory variables.

8.3 THE MULTIPLE REGRESSION MODEL

The examples presented in the previous sections illustrate a simple linear additive model between two variables. In this, the behaviour of the response is assumed to be predictable given a linear additive function of fixed explanatory information. If the hypothesised model is correct, the model will yield a high value for r^2 and generate an essentially random pattern of residuals.

The 'real world' is rarely so neat. Most social and environmental systems display a more complex assortment of relationships. These may involve many different responses and explanatory variables, a variety of linear and non-linear functional arrangements, and be compounded by dynamic linkages which multiply or dampen the relationship depending on context. This array of complexity corresponds to the area of statistics known as multivariate analysis. Many of the procedures developed here are used in geography for classification, data management and screening, and the reduction of complex information to value-rich variables (a form of data puréeing).

The multiple regression model is an extension of the simple linear regression model of section 8.2 which may be applied when the behaviour of a single response variable is linked to several explanatory variables. Taylor (1980) describes how this model may be implemented using rainfall data gathered in California.

8.3.1   Fitting a multiple regression model in GLIM

The data examined in Taylor (1980) consist of a series of observations collected at thirty Californian weather stations and include:

1 PPT   annual average precipitation (inches).
2 ALT   altitude of the weather station in feet above sea level.
3 LAT   latitude of the weather station in degrees.
4 DIST   straight-line distance from each weather station to the coast (miles).

These stations were dispersed throughout the State and include coastal and inland sites, highland and lowland sites, and northerly and southerly sites (Figure 8.7(a)). For the purposes of his example, PPT was treated as the response variable, and the three others as explanatory variables.

Regression is a plausible technique for describing the patterns in the observed data because all four variables are measured as continuities (though PPT, LAT and DIST cannot be less than 0), and it can be argued that the relationship between PPT and the others is likely to be additive. Given this, the following hypotheses are worth testing:

1 PPT increases with altitude (due to the effect of topography).
2 PPT increases towards the north of the State (as northerly weather stations are more likely to be affected by westerly weather systems).
3 PPT decreases farther inland.

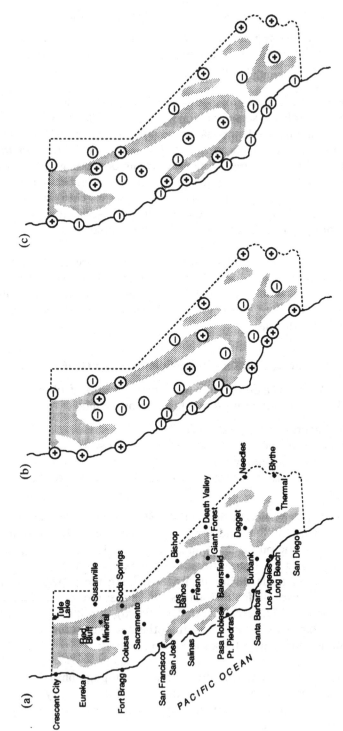

Figure 8.7 (a)Location of Californian weather stations; (b) Map of regression residuals; (c) Map of regression residuals from second analysis

The GLIM commands needed to specify the multiple regression model for these data are set out in Printout 8.3, which is a logged transcription of the computing session. Notice that $UNITS has been set to 30 to accommodate the data for the thirty weather stations.

The four variables are defined in the $DATA command, and are read into GLIM from a secondary data file. Instead of using $READ, which permits data entry directly from a terminal, the data values have already been typed into a file termed 'taylor'. To read these into GLIM the $DINPUT command is used. This is followed by the channel number for data entry, which in this case is channel 1. (Notice that this channel number has been attached to data file 'taylor' as a parameter field on the first line of the transcript – the RUN command used by the computer operating system.) PPT is defined as the response using $YVAR, and the identity link and Normal error are defined as before.

The total sum of squares associated with these data (8,012) is calculated from the first FIT command which fits %GM to the model. The second fit, $FIT ALT+LAT+DIST, fits the three explanatory variables to the model, reducing deviance to 3,202 for the loss of three degrees of freedom (29–26). Two options are selected with the display command: E for regression parameter estimates, and R for regression residuals. The maximum likelihood estimates of this model are:

$$PPT = -102.4 + 0.004ALT + 3.451LAT - 0.143DIST \qquad (8.16)$$
$$ (29.2) \qquad (0.001) \qquad (0.79) \qquad (0.036)$$

(estimated standard errors for the parameters are given in brackets). These suggest that if altitude increases by one unit, that is, one foot, PPT will increase by 0.004 inches. Similarly, a one degree move north suggests PPT should increase by 3.451 inches, and a move of one mile inland from the coast should decrease PPT by 0.143 inches. The intercept value of -102.4 appears to suggest that for values of ALT, LAT and DIST of 0 – corresponding to sea level, coastal sites situated on the equator – the value of PPT will be -104.2 inches. This is clearly peculiar, not only because negative precipitation is not meaningful, but also because the absolute location of California on the globe renders such values for LAT impossible. This suggests that the intercept is not interpretable in this context. As a general rule, intercepts should not be interpreted in social or environmental regression models unless theory requires it, or unless there are observed values of the response variable associated with values of the explanatory variables which are near 0.

The individual t statistics for these parameters (calculated by dividing the parameter estimate by its estimated standard error) are -3.51 (intercept), 3.36 (ALT), 4.34 (LAT), -3.93 (DIST), and all are significant at the 5 per cent significance level (two-tailed test). Thus the results of the analysis appear to confirm the hypotheses listed above. However, though

Printout 8.3 GLIM analysis of Taylor's Californian rainfall data

```
£r *glim 1=taylor
  $UNITS 30
  $DATA PPT ALT LAT DIST
  $DINPUT 1$
```

$LOOK	PPT	ALT	LAT	DIST$
1	39.57	43.00	40.80	1.000
2	23.27	341.0	40.20	97.00
3	18.20	4152.	33.80	70.00
4	37.48	74.00	39.40	1.000
5	49.26	6752.	39.30	150.0
6	21.82	52.00	37.80	5.000
7	18.07	25.00	38.50	80.00
8	14.17	95.00	37.40	28.00
9	42.63	6360.	36.60	145.0
10	13.85	74.00	36.70	12.00
11	9.440	331.0	36.70	114.0
12	19.33	57.00	35.70	1.000
13	15.67	740.0	35.70	31.00
14	6.000	489.0	35.40	75.00
15	5.730	4108.	37.30	198.0
16	47.82	4850.	40.40	142.0
17	17.95	120.0	34.40	1.000
18	18.20	4152.	40.30	198.0
19	10.03	4036.	41.90	140.0
20	4.630	913.0	34.80	192.0
21	14.74	699.0	34.20	47.00
22	15.02	312.0	34.10	16.00
23	12.36	50.00	33.80	12.00
24	8.260	125.0	37.80	74.00
25	4.050	268.0	33.60	155.0
26	9.940	19.00	32.70	5.000
27	4.250	2105.	34.09	85.00
28	1.660	−178.0	36.50	194.0
29	74.87	35.00	41.70	1.000
30	15.95	60.00	39.20	91.00

```
  $YVAR PPT$
  $ERR N$
  $LINK I$
  $FIT$
```

CYCLE	DEVIANCE	DF
1	8012.	29

```
  $FIT ALT+LAT+DIST$
```

CYCLE	DEVIANCE	DF
1	3202.	26

```
  $D ER$
```

	ESTIMATE	S.E.	PARAMETER
1	−102.4	29.20	%GM
2	0.4091E-02	0.1218E-02	ALT

	3	3.451	0.7947	LAT
	4	−0.1429	0.3634E−01	DIST
SCALE PARAMETER TAKEN AS				123.1

UNIT	OBSERVED	FITTED	RESIDUAL
1	39.57	38.48	1.090
2	23.27	23.91	−0.6439
3	18.20	21.28	−3.075
4	37.48	33.78	3.704
5	49.26	39.46	9.797
6	21.82	27.59	−5.773
7	18.07	19.18	−1.113
8	14.17	23.10	−8.932
9	42.63	29.26	13.37
10	13.85	22.89	−9.037
11	9.440	9.366	0.7381E−01
12	19.33	20.94	−1.608
13	15.67	19.45	−3.776
14	6.000	11.10	−5.098
15	5.730	14.89	−9.157
16	47.82	36.62	11.20
17	17.95	16.71	1.241
18	18.20	25.42	−7.220
19	10.03	38.75	−28.72
20	4.630	−5.953	10.58
21	14.74	11.82	2.924
22	15.02	14.32	0.7036
23	12.36	12.78	−0.4207
24	8.260	18.03	−9.774
25	4.050	−7.447	11.50
26	9.940	9.858	0.8202E−01
27	4.250	11.76	−7.509
28	1.660	−4.835	6.495
29	74.87	41.55	33.32
30	15.95	20.17	−4.221

$STOP$

Note: Commands prefixed by £ are operating system commands
Commands prefixed by $ are GLIM commands

these individual components appear to fit the data satisfactorily, the overall performance of the model is poor. The ratio of the deviance of fit 2 to fit 1 (3,202/8,012) is 0.4. This corresponds to the error sum of squares in a model estimated using ordinary least squares. Consequently, the value of r^2 for this model is $1-ESS = 0.60$, indicating that only 60 per cent of the variability in the observed values of PPT have been described by the model.

8.3.2 Residual analysis in multiple regression

The analysis of residuals is of great importance in assessing the fit of a multiple regression model. A number of different types of residual plot

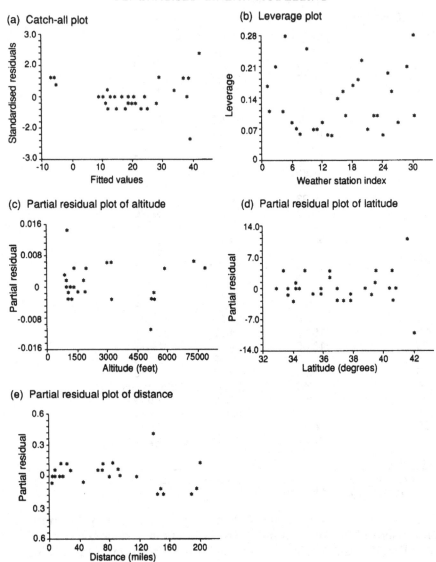

Figure 8.8 Graphical plots of multiple regression analysis of Californian rainfall data

are illustrated in Figure 8.8. Plots (a) and (b) are catch-all and leverage plots, created in exactly the same way as for the unemployment data in section 8.2. A quick inspection of plot (a) suggests that though the linear functional form is reasonable, some transformation might be useful. Two of the weather stations also appear to be outliers with standardised residuals exceeding 2 (stations 19 and 29). These are the two most northerly weather stations in the sample, and may be affected by weather systems which are somewhat different from the rest of California. The critical value for the leverage plot is $3 \times 4/30 = 0.4$. None of the leverage values exceed this critical value so we can assume that there are no particularly influential observations in the data set.

Plots (c) to (e) are partial residual plots. These are created by multiplying each observed value of an independent variable by its parameter estimate and adding this to the ordinary least squares residuals from the model:

$$e_p = e_i + \hat{b}x_i \qquad (8.17)$$

for $j = 1, \ldots, k$ independent variables. The purpose of these plots is to assess the relationship between the response variable and each of the independent variables whilst controlling for the effects of the other independent variables. This information is needed because the independent variables in a multiple regression model potentially affect the response in two distinct ways: (a) as a series of independent effects, (b) as a series of interactions which reflect the fact that the independent variables may be interrelated themselves. The latter feature is termed collinearity. If it is particularly severe, it may seriously affect the performance of the model, leading to misinterpretation of the estimates produced. (For details of multicollinearity – collinearity between many independent variables – see Hoerl and Kennard 1970, Brown and Zidek 1980 and Gherardini 1980.)

The shape of the partial residual plots indicates whether there are model misspecifications associated with a specific independent variable. If the shape is essentially linear in form then the model would seem to be adequate. However, if the shape is curvilinear, diamond or wedge shaped, or if there are stray points, these suggest misspecifications due to a poor functional form, heteroscedasticity or outliers. In plot 8.8(c) there are clearly two outlying values associated with partial residuals exceeding plus or minus 0.008. These are associated with the two most northerly weather stations and add extra weight to the suspicion that they behave differently from the rest of the sample. Apart from these, the remaining points are sufficiently linear in shape to support the analysis at this stage. The same story emerges from plots 8.8(d) and 8.8(e), where the peculiarity of weather stations 19 and 29 is again confirmed.

8.3.3 Introducing binary dummy explanatory variables

One of the possible reasons for the relatively poor fit of the Californian precipitation model may be due to omitted explanatory variables. The scattered distribution of the weather stations across the State has only been incorporated in the model in terms of the DIST and LAT variables. However, as some of these stations are situated in the lee of mountains whilst others are on the windward side, it seems reasonable that a rain-shadow effect may be important. Further support may be obtained for this by dividing the weather stations into two groups – windward and leeward sites – and comparing the signs of the residuals associated with each (Table 8.4). There is a clear difference between the two groups with windward sites mainly displaying positive residuals, whilst residuals to leeward are generally negative. This becomes even clearer when the residuals are displayed spatially (Figure 8.7(b)).

This clear spatial structure to the residuals can be incorporated explicitly within the regression model by creating a binary dummy variable (RS) to represent the rain-shadow effect. The values associated with this categorical independent variable are 1, if the site of the weather station is leewards, and 0, if the site is windward. An hypothesis associated with this variable is that precipitation is less at weather stations which are leeward.

Table 8.4 Residuals from Taylor's Californian regression

Station	Site	SR
Eureka	W	0.10778
Red Bluff	L	−0.06175
Thermal	L	−0.31255
Fort Bragg	W	0.35549
Soda Springs	W	1.03547
San Francisco	W	−0.54242
Sacramento	L	−0.10417
San Jose	L	−0.82885
Gian Forest	W	1.39556
Salinas	L	−0.84242
Fresno	L	0.00691
Pt. Piedras	W	−0.15107
Pasa Robles	L	−0.35022
Bakersfield	L	−0.47163
Bishop	L	−0.89127
Mineral	W	1.09891
Santa Barbara	W	0.11793
Susanville	L	−0.71400
Tule Lake	L	−2.85008
Needles	L	1.08086
Burbank	W	0.27401
Los Angeles	W	0.06664
Long Beach	W	−0.04008
Los Banos	L	−0.90640
Blythe	L	1.15750
San Diego	W	0.00801
Daggett	L	−0.70884
Death Valley	L	0.68843
Crescent City	W	3.38405
Colusa	L	−0.39953

Note: L: leeward sites; W: windward sites

The effect of adding the four explanatory variables to the data is to reduce deviance from 8,012 for 29 degrees of freedom to 2,098 for 25 degrees of freedom. The difference, 5,914 for 4 degrees of freedom, represents the overall effect of the extended model. In terms of r^2 the new model accounts for 5,914/8,012 = 0.738, or about 74 per cent of the variability in PPT, a considerable improvement over the original model.

The parameter estimates and standard errors (in brackets) for this model are:

$$PPT = -97.88 + 0.002ALT + 3.45LAT - 0.05DIST - 15.85RS$$
$$ (21.13) \quad (0.001) \quad\quad (0.66) \quad\quad (0.039) \quad\quad (4.37)$$

$$(8.18)$$

Once again the hypotheses suggested for ALT, LAT and DIST are supported by this model. However, some care needs to be taken in interpreting the parameter for RS. Given its values of 1 or 0, this parameter will only be included in the model for weather stations which are located leewards. Thus predicted values for leeward stations will be based on four explanatory variables whereas those for windward stations will be based on three. The dummy thus represents the difference in the predicted value of PPT of being a leeward rather than a windward weather station. The sign of the parameter suggests that being leeward reduces expected precipitation by 15.85 inches, compared with windward stations having identical values for ALT, LAT and DIST.

As with the first analysis it is important to examine the residuals from this extended model to see if they form any type of pattern (Figure 8.7(c)). The partial residual plots show that the two extreme weather stations noted in the earlier analysis still remain relatively extreme. This suggests that the inclusion of the rain-shadow dummy has done little to help incorporate their information into the body of the model.

There are a number of things which might be done to try to improve the fit of the model. The simplest is to eliminate the effects of weather stations 19 and 29 from the model either by removing them from the data set explicitly, for example by deleting their data values, or by weighting them so that their effect is nullified. By declaring that they have a weight of 0, and including in the model only those observations with a non-zero weight, the information collected for them may be eliminated from the fitted model. This may be set up with ease in GLIM using the $WEIGHT command. To do this, a variable (W) containing the weights of each of the observations is set up using $CALCULATE. Each observation is given the weight of 1. Having done this, the $EDIT command is used to reset the weights of observations 19 and 29 to 0. In order to fit the model excluding 19 and 29, W is declared as a weight, and a $FIT command is issued. GLIM fits the required effects subject to the conditions set by the weight. The full sequence is:

217

```
$CALCULATE W=1
$EDIT 19 W 0
$EDIT 29 W 0
$WEIGHT W
$FIT ALT+LAT+DIST+RS$
```

The effect of this is to increase r^2 for the twenty-eight weather stations to 89 per cent. (The equivalent manipulations in MINITAB make use of the OMIT command.)

However, a comparison with the r^2 value from the previous model is not entirely appropriate because the number of independent items of data being fitted is different. A modified r^2 value which scales r^2 by its degrees of freedom should be used instead. Instead of dividing the error sum of squares by $n-1$, the term $n-p$ is used instead, where p refers to the number of parameters being fitted. This modification provides for a more effective comparison between models, and shows that the fit improves from an r^2 of 70 per cent for the model with the extreme values included, to 87 per cent with them excluded.

8.3.4 Some further modifications

The examination of the residuals from the fitted models provides an effective way of improving the description of the patterns in the observed response variable. Jones (1984) notes that an r^2 of 90 per cent may be achieved for these data by further modification. By inspecting the partial residual plots from the model excluding the two most northerly sites, he notes that stations 15 and 18 are potential outliers from the modified model. These appear to be more extreme in the plots produced for distance and the rain shadow suggesting that these variables may interact in some way. A variety of different terms might be tried to reproduce the assumed interaction. The one suggested by Jones is to multiply the values of DIST by RS to create a new variable (INT) which is added to the model.

In addition to creating this interaction, Jones notes that the partial residual plot between precipitation and altitude indicates that their relationship is not linear. As the relationship between precipitation and the remaining independent variables does appear to be linear, a transformation of the altitude variable seems called for. A logarithmic transformation is not possible because of the negative values associated with one of the weather stations, one located below sea level in Death Valley. However, this problem may be removed by adding a constant to each of the values of altitude so 'raising' Death Valley above sea level. Using a constant of 500 and logarithms (base 10) the final model proposed by Jones (1984) is:

$$\text{PPT} = -108 + 7.9\text{ALT} + 2.9\text{LAT} + 0.1\text{DIST} - 6.9\text{RS} - 0.12\text{INT}$$
$$(16.5) \qquad (3.1) \qquad (0.4) \qquad (0.03) \qquad (3.7) \qquad (0.03)$$

(standard errors in brackets). $\qquad\qquad\qquad\qquad\qquad\qquad\qquad$ (8.19)

218

This model may be interpreted as follows. Precipitation is linearly related to latitude, distance from the coast and a modified measure of altitude. This modification uses a logarithmic transformation and a constant and suggests that precipitation initially increases rapidly with altitude, but that this effect becomes less marked at higher levels. The rain shadow is seen to be important but is more particularly relevant if the site concerned is also far inland.

8.3.5 Introducing multi-way dummy explanatory variables

The use of dummy variables to examine qualitative differences among continuous measurements is frequently a useful extension to a regression analysis. However, care must be taken in specifying and interpreting such dummies. In the previous section the effect of the rain-shadow dummy was introduced in two distinct ways. First, by including it as an additive parameter (RS). Second, by multiplying it with the distance variable to create the INT interaction. The final model included both RS and INT. It would have been perfectly reasonable to have eliminated the RS effect once INT was present, thus revealing a third possible way of incorporating the dummy.

The three distinct ways of handling dummy variables need to be distinguished because they affect the model in different ways. The presence of RS may be gauged by noting that the effect of being leeward reduces the value of the intercept for the model by approximately 16 inches. The slope coefficients for leeward and windward sites are identical. If INT were present but RS were removed, the intercepts for leeward and windward models would be identical but the slope coefficients for distance would differ. In the third model in which both effects are present in the model, both the intercepts and slope coefficients of the model will differ between leeward and windward sites. (For further details about this topic, see Chatterjee and Price 1977, Pindyck and Rubinfeld 1976: section 3.8, and Wrigley 1985: 91–4.)

It is perfectly reasonable to use dummy variables to incorporate qualitative independent variable effects which are multi-level rather than binary. However, the multi-level dummy variable has to be treated somewhat differently. The correct procedure for handling it is to re-express the information it contains as a series of binary dummies. Thus the three levels of the dummy variable HOUSING:

0 corresponding to owner occupation,
1 corresponding to private renting,
2 corresponding to council renting,

should be recoded into two binary dummies, $H1$ and $H2$, where for the former, 0 represents owner occupation and 1 private renting, and for $H2$,

0 represents owner occupation and 1 council renting. (In general, if there are n levels in the multi-way dummy, a series of $n-1$ binary dummies will be needed to reproduce the observed data correctly.) When fitted in a regression model, $H1$ represents the effect on the response variable of being a private renter rather than an owner occupier, whereas $H2$ represents the effect on the response of being a council tenant rather than an owner occupier. The two terms could be included in the model as additive effects, interactions or both.

8.4   THE ANALYSIS OF VARIANCE

The analysis of variance is a statistical technique which is particularly useful if the data to be analysed are continuous but have been allocated to non-overlapping categories by some classification scheme. Such procedures are common in the social and environmental sciences, and are frequently developed into experimental designs of varying degrees of complexity. (A discussion of experimental designs is presented in Chapter 10.) The motivation for the analysis of variance, also sometimes termed ANOVA, is exactly the same as for regression in that the variability in a response is described using explanatory variables. Though regression and ANOVA look different, they are formally identical. The differences between them arise because of the way the variation in the response variable is partitioned into recognisable subsets of explanatory information.

8.4.1   One-way ANOVA

Table 8.5 is an example of a one-way analysis of variance table. This has been created by measuring male economic activity rates in three samples of wards in County Durham. The data come from the 1981 Population Census collections held in NOMIS. Three distinct types of ward have been identified based on the Office of Population Censuses and Surveys (OPCS) urban–rural classification. These are:

1 wards which are wholly urban,
2 wards which are wholly rural,
3 wards which are 'mixed'.

Each classification contains twenty wards. Table 8.5 is said to be a one-way classification because only one categorical dimension has been used: the urban–rural continuum.

The object of analysis of variance is to test whether the differences between the means of the different classifications are significant. In effect, this tests whether male economic activity rates depend on ward type. The technique involves:

Table 8.5 Male economic activity rates in samples
of wards in County Durham

Ward	Urban	Rural	Mixed
1	80.26	76.54	80.00
2	76.73	79.60	74.97
3	72.15	58.65	78.67
4	74.00	82.28	73.51
5	80.32	77.48	78.22
6	80.73	71.80	87.05
7	85.98	74.77	89.40
8	76.13	69.93	77.81
9	77.93	74.31	79.57
10	75.87	80.37	80.76
11	78.93	74.38	74.81
12	73.81	80.81	76.92
13	79.18	76.82	70.41
14	78.04	75.29	72.88
15	81.47	80.00	72.91
16	82.57	78.26	72.42
17	84.21	69.67	75.08
18	73.78	73.90	75.08
19	79.87	77.96	72.93
20	79.23	76.85	69.53

1 testing the null hypothesis that the samples have been drawn from the same population (or from different populations with equal means), and
2 comparing the sample means.

In its simplest form, analysis of variance involves partitioning the total amount of variability (total sum of squares) of a data set to allow two estimates of the overall variance of the data to be calculated. As was shown in Part I, the overall variance consists of a measure of the variability of all the observations in the data set around the overall mean. The first estimate of this may be produced by calculating the values of the variances in each of the classifications, that is, describing the variability of the wholly urban, wholly rural and mixed wards around their three respective means. The second estimate may be produced by calculating the variance of these three sample means around the overall mean. These two distinct estimates are termed:

1 The within variance, referring to variability within each classification.
2 The between variance, referring to the variability between the three classifications.

If the null hypothesis is true, the ratio of the two should be unity. That is:

$$\frac{\text{Between classification estimate}}{\text{Within classification estimate}} = 1 \qquad (8.20)$$

If the null hypothesis is not true, then it is likely that the between-classification estimate will be greater than the within-classification estimate, indicating that the three classifications refer to different populations with different mean parameters.

Because the observed data are generated from a sample (or are to be treated as a sample), it is possible that a value for the ratio other than unity may occur even though the classifications come from the same population, or different populations with equal means. It is therefore necessary to compare the observed value for the ratio with a value which might be expected to have arisen purely by chance. The F distribution (see Chapter 5) is used to provide the necessary bench-mark against which to make the assessment.

8.4.2   GLIM and MINITAB analyses of Table 8.5

The basic idea behind analysis of variance can be illustrated by calculating the value of the variance–estimates ratio for the data in Table 8.5. Printout 8.4 presents the GLIM analysis of these data; Printout 8.5 the equivalent MINITAB analysis.

The structure of these transcriptions is very similar to those associated with the linear regression models. In MINITAB, the data are read into three column vectors, C1–C3, where C1 contains the data for the wholly-urban wards, C2 the wholly-rural wards, and C3 the mixed wards. The command AOVONEWAY is used to produce the output. MINITAB produces tabular output in response to this command. This consists of an incomplete three-row, four-column matrix containing details on the number of degrees of freedom, sums of squares and mean squares associated with the model and the errors. Row 3 displays the total sum of squares in the data. This is 1,357.3 for 59 degrees of freedom. By regressing the observed economic activity rates against their ward classification, the total sum of squares is reduced to 1,260.8 for 57 degrees of freedom. Thus the model fitted to the data reduces total sum of squares by 96.5 for 2 degrees of freedom. Line 1 is thus the equivalent of the regression sum of squares, and line 2 the error sum of squares.

Columns 3 and 4 of the matrix contain the information needed to test the null hypothesis. Column 3 contains mean square errors. These are produced by dividing the figures in column 2 by the degrees of freedom in column 1. The mean square error for line 1 represents the estimate of the variance produced from between the classifications. The within-classification estimate is the mean square error from line 2. The ratio of line 1 to 2 is displayed in column 4. This is the calculated value of the F statistic for 2 and 57 degrees of freedom. This may be checked for significance using tables of the F distribution, where the 2 subscript refers to the column of the table, and the 57 subscript to the row. The value in the table associated

Printout 8.4 GLIM analysis of Table 8.5 data

```
£r * glim 1=−a
  $UNIT 60
  $DATA EA
  $DINPUT 1$
  $FAC WT 3
  $CALC WT=%GL(3,20)
  $YVAR EA$
  $FIT$
```

CYCLE	DEVIANCE	DF
1	1357.	59

```
  $D E$
```

	ESTIMATE	S.E.	PARAMETER
1	76.90	0.6192	%GM
	SCALE PARAMETER TAKEN AS		23.01

```
  $FIT WT$
```

CYCLE	DEVIANCE	DF
1	1261.	57

```
  $D E$
```

	ESTIMATE	S.E.	PARAMETER
1	78.56	1.052	%GM
2	−3.076	1.487	WT(2)
3	−1.913	1.487	WT(3)
	SCALE PARAMETER TAKEN AS		22.12

```
  $STOP$
```

Note: Commands prefixed by £ are operating system commands
Commands prefixed by $ are GLIM commands

with these subscripts (3.15 at the 95 per cent level, 4.9 at the 99 per cent level) is the value of F which may be expected to occur by chance. As the calculated value is less than either of these tabulated values we cannot reject the null hypothesis of no difference. In other words, there is no evidence to suggest that male economic activity rates differ according to ward type.

The GLIM analysis of these data confirm the MINITAB findings. However, there are a number of important distinctions which need to be noted in comparing the two transcriptions:

1 Only one data item (EA) is read into GLIM. This gives the observed rate of male economic activity in the 60 wards (this will be used as the response variable).
2 Information on ward-type (that is, wholly urban, wholly rural, or mixed) which is to be used as the explanatory information is generated internally by GLIM using the $FACTOR and $CALCULATE commands.

In MINITAB the observed rates for the three types of ward are stored in

Printout 8.5 MINITAB analysis of Table 8.5 data

```
£R *MINITAB
-READ 'URT1' C1
      20 ROWS READ
   C1
    80.26  76.73  72.15  74 . . .

-READ 'URT6' C2
      20 ROWS READ
   C2
    76.54  79.60  58.65  82.28 . . .

-READ 'URT25' C3
      20 ROWS READ
   C3
    80.00  74.97  78.67  73.51 . . .

-PRINT C1-C3
   ROW        C1        C2        C3
     1      80.26     76.54     80.00
     2      76.73     79.60     74.97
     3      72.15     58.65     78.67
     4      74.00     82.28     73.51
     5      80.32     77.48     78.22
     6      80.73     71.80     87.05
     7      85.98     74.77     89.40
     8      76.13     69.93     77.81
     9      77.93     74.31     79.57
    10      75.87     80.37     80.76
    11      78.93     74.38     74.81
    12      73.81     80.81     76.92
    13      79.18     76.82     70.41
    14      78.04     75.29     72.88
    15      81.47     80.00     72.91
    16      82.57     78.26     72.42
    17      84.21     69.67     75.08
    18      73.78     73.90     75.08
    19      79.87     77.96     72.93
    20      79.23     76.85     69.53

-DESCRIBE C1-C3
```

	N	MEAN	MEDIAN	TRMEAN	STDEV	SEMEAN
C1	20	78.559	79.055	78.503	3.623	0.810
C2	20	75.48	76.68	76.04	5.26	1.18
C3	20	76.65	75.08	76.33	5.06	1.13

	MIN	MAX	Q1	Q3
C1	72.150	85.980	75.935	80.627
C2	58.65	82.28	74.00	79.26
C3	69.53	89.40	72.91	79.34

```
AOVONEWAY C1-C3
ANALYSIS OF VARIANCE
```

SOURCE	DF	SS	MS	F
FACTOR	2	96.5	48.2	2.18
ERROR	57	1260.8	22.1	
TOTAL	59	1357.3		

INDIVIDUAL 95 PCT CI'S FOR MEAN
BASED ON POOLED STDEV

LEVEL	N	MEAN	STDEV				
C1	20	78.559	3.623			(------------)	
C2	20	75.483	5.260	(-------*---------)			
C3	20	76.646	5.057	(----------*-------)			

POOLED STDEV = 4.703

```
                          74.0      76.0      78.0      80.0
```

−STOP

Note: Commands prefixed by £ are operating system commands
Commands prefixed by − are MINITAB commands

three distinct external data files – URT1, URT6, URT25. These are read separately, copying their data in C1, C2 and C3. The analysis of variance is performed merely by identifying the columns concerned. In GLIM, all the data are read in from a single external file, –a. In order to reproduce the structure of the ward classification, a three-level categorical variable, WT, is defined and observed economic activity rates in EA related to it using $CALCULATE. Level 1 of WT corresponds to column 1 of Table 8.5 (wholly-urban wards), level 2 to wholly-rural wards, and level 3 to mixed urban–rural wards. The $CALCULATE command works by generating the levels of the dummy variable WT internally. A system function (%GL) is used to generate the levels associated with each observed value. Though the command looks rather complex, its syntax is fairly straightforward. It works as follows. Within the brackets following the %GL terms are two numbers (3,20). The first number represents the number of levels specified for WT in the $FACTOR command. The second figure represents the levels of WT which are associated with the observed items of data in EA. GLIM generates factor levels for WT which run from 1 to 3 in sequences of 20, that is:

1 1 1 1 1 1 1 1 1 1 1 1 1 1 1 1 1 1 1 1 2 2 2 . . .

This sequence continues until all 60 levels have been generated, this upper limit being set automatically to the value of the $UNITS command. Thus the first twenty observed values in EA are set at level 1 of WT, indicating that they are wholly-urban wards. The second set of twenty observed values is set at level 2, the third at level 3. (Notice this coding corresponds to reading the data from Table 8.5 a column at a time.) It is important to note that the values given in the %GL command reflect the order in which the data are to be entered. If the observed data were to be entered a row at a time, the appropriate coding for %GL would be %GL(3,1), generating sequences of

1 2 3 1 2 3 1 2 3 1 2 3 . . .

The remaining GLIM commands are as for the regression examples, with $YVAR being defined as EA, and the $ERR and $LINK being Normal and Identity respectively. The first FIT command fits the grand mean effect to the observed data. Its value of 1,357 for 59 degrees of freedom is identical to the total sum of squares produced by MINITAB. The second $FIT command fits the main effects of the ward classification. The value of deviance associated with this is 1,261 for 57 degrees of freedom, a value which is equivalent to the error sum of squares in MINITAB. The difference is the effect of deviance of fitting the WT term to the first model. This is identical to the sum of squares associated with line 2 of the MINITAB analysis of variance table. The scale parameter contains the mean square errors associated with the two models, and so can be used to calculate the F ratio test.

One important point of difference between the two programs should be noted. The values for the column means generated by MINITAB appear to differ from those produced by GLIM. In MINITAB, the mean rate for male economic activity in wholly-urban wards is 78.6, in wholly-rural wards 75.5, and in mixed wards 76.7. The GLIM equivalents are produced in response to the $DISPLAY command. The estimate for %GM is 78.6, equivalent to that for the wholly-urban wards, but the estimates for wholly-rural wards – WT(2) – and mixed wards – WT(3) – are small negative numbers. In fact, if WT(2) is added to %GM, the difference is identical to the value produced by MINITAB for wholly-rural wards. Similarly, the values for mixed wards are also identical if WT(3) is added to %GM. Both programs produce exactly the same information but record it differently.

The detailed reason for this difference will be explained in Chapter 9 in connection with the constraints on log-linear models. Constraints are used in order to produce parameter estimates. They are not important in themselves and in no way affect the overall performance of the model. This is clear from the examples because the values of the sums of squares and mean squares are identical. However, they do affect the form of the parameter estimates, hence the differences between these. MINITAB uses a form of constraint termed the usual constraints, a form of centre-effect coding in which the sum of the effects in the single explanatory variable (WT) is assumed to equal zero. In contrast, GLIM uses a form of constraint in which the mean economic activity rate of wholly-urban wards is treated as a baseline. The mean rates for the other ward classifications are thus defined with respect to this rather than an overall mean. Once this is recognised it is possible to interpret the two procedures consistently.

8.4.3 Assumptions and residuals

Like the regression model there is a series of assumptions underlying the use of the analysis of variance. The two most important of these are:

1 that the predictable mean of the response is the sum of a series of effects which are linear in their parameters, and

2 that the error terms are uncorrelated and have constant variance.

These are equivalent to the assumptions used in regression. It is also frequently asserted that the errors must be derived from a Normal distribution, but as with other models in the exponential family, the assumption of a specific probability process is not a fundamental requirement (McCullagh and Nelder 1983). The three samples are also assumed to have been drawn by some form of independent selection process and have identical population standard deviations. This latter assumption is what allows us to draw two distinct estimates of the variance.

The procedures used for assessing the acceptability of these assumptions are also similar to those used with regression. Standardised residuals may be produced within GLIM and plotted against fitted values to identify specific cells of the table which are extreme in some way. Similarly, the residuals may be plotted spatially to see if clusters are evident, indicative of spatial autocorrelation. Finally, leverage statistics may be produced to identify unusual cells, columns or rows whose influence is extreme. The latter check is particularly important in tabular analyses where the main sources of the variability may arise because of unusually important subsets of the table.

8.4.4 Higher-order analysis of variance

The general motivation for analysis of variance has been set out using the one-way design. However, it is more likely that observed data may be subject to categorisation by more than one type of classification. This leads to the need to consider higher-order, multi-way analyses of variance, such as frequently occur in the biological, medical and psychological sciences.

Table 8.6 contains an extract from Table 16.6 of Blalock (1979) in which a series of murder rates have been cross-classified by a three-level city-type variable and a binary region variable. The cells of the table are measured as continuous variables, but the explanatory structure of the table is categorical. The analysis of this table is identical to that for Table 8.5 in that the observed murder rates are defined as the response variable and city type and region are defined as explanatory variables.

Two features distinguish this table from Table 8.5. First, for each combination of the explanatory variables there are four replications – four distinct observations on the murder rate. Second, the total sum of squares can be partitioned into a wider range of possible sources. In addition to effects due to city type, region and errors, we may also define an interaction between them both. As with the interaction presented in connection with the precipitation data, this term is used to refer to the way

Table 8.6 Two-way analysis of variance data

		CITY TYPE					
		I		T		G	
	NE	4.3	5.9	5.1	3.6	3.1	3.8
		2.8	7.7	1.8	3.3	1.6	1.9
REGION							
	SE	12.3	9.1	6.2	4.1	6.2	11.4
		16.3	10.2	9.5	11.2	7.1	12.5

Note: Region: NE = North-east, SE = South-east
City type: I = Industrial, T = Trade, G = government

Source: After Blalock (1979)

murder rates may depend not just on city type or region, but on their combined effect. Thus murder rates may be higher in north-eastern industrial cities simply because of the special combination of effects operating there. These rates would be higher than those expected simply by looking at the north-east without distinguishing between cities, or by looking at industrial cities without distinguishing between regions.

Blalock (1979: 359) notes that this modification leads to three distinct types of hypothesis which need testing:

1 the population column means are equal;
2 the population row means are equal;
3 the population is additive.

The third of these is a test for the presence of the interaction effect. Printout 8.6 summarises the analysis of these data.

The total sum of squares for these data is estimated from the FIT command as 373.5 for 23 degrees of freedom. By fitting the term CT (city type) to the base model, the total sum of squares falls by 42.3 for 2 degrees of freedom. This corresponds to the between columns estimate. By fitting the term R (region) to the base model, the total sum of squares falls by 211.2 for 1 degree of freedom. This corresponds to the between rows estimate. Their combined effect is to reduce the deviance value by about 250 for 3 degrees of freedom. The interaction effect may be fitted to the model containing the two effects using GLIM's dot notation (CT.R). The addition of this reduces deviance by a further 8 for 2 degrees of freedom. The combined effect of the two main effects – the row and column effects – and the interaction is to reduce deviance by 260 for 5 degrees of freedom. This means that the variability attributable to the errors is 112 for 18 degrees of freedom.

The mean square errors associated with these effects are:

between columns – 42.3/2 = 21.2
between rows – 211.2/1 = 211.2

Printout 8.6 GLIM analysis of Table 8.6 data

```
$UNITS 24
$DATA N
$DINPUT 1$
$FAC R 2 CT 3
$CALC CT=%GL(3,2):R=%GL(2,2)
$YVAR N
$FIT$
```

CYCLE	DEVIANCE	DF
1	373.5	23

```
$FIT R$
```

CYCLE	DEVIANCE	DF
1	162.3	22

```
$FIT CT$
```

CYCLE	DEVIANCE	DF
1	331.2	21

```
$FIT +R$
```

CYCLE	DEVIANCE	DF
1	120.0	20

```
$FIT +R.CT$
```

CYCLE	DEVIANCE	DF
1	112.0	18

```
$CALC 373.5–112$
   261.5

$CALC 373.5–162.3$
   211.2

$CALC 373.5–331.2$
   42.30

$STOP
```

Note Commands prefixed by £ are operating system commands
Commands prefixed by $ are GLIM commands

interaction – 8/2 = 4
errors – 112/18 = 6.2

The interaction effect is tested first by forming an F ratio using the interaction mean square error as the numerator and the error mean square error as the denominator. This gives a value of 0.644. As this is less than unity we can assume that there is no evidence for an interaction. The small reduction in deviance associated with the interaction effect may thus be assumed to be the result of sampling fluctuations rather than a real effect.

Blalock shows that having eliminated the interaction as a real effect, its deviance value can be incorporated into the error component, and new F ratios calculated for the two main effects. The new error sum of squares rises from 112 for 18 degrees of freedom to 120 for 20 degrees of freedom,

229

producing a new error mean square of 6. The two F ratios using 6 as a denominator are:

between columns $- 21.2/6 = 3.53$
between rows $\quad - 211.2/6 = 35.2$

Tests for these ratios may be made using F tables with 2 and 20 degrees of freedom for the between columns estimate, and 1 and 20 degrees of freedom for the between rows estimate. Both of these ratios are significant at the 55 per cent significance level. This means that there are clear relationships between both region and murder rates and city type and murder rates. Blalock (1979: 362) suggests that

> when we control for region by letting this factor explain all it can of the variation in murder rates and then letting city type explain what it can of the remainder, we now get a significant relationship between city type and murder rates.

This contrasts with results produced for an analysis of variance table in which region was not identified. This suggested that murder rate was not affected by city type. (For further details of this example or analysis of variance in general, see Blalock 1979.)

8.4.5 ANOVA with random effects: components of variance

The analyses presented in the previous sections assume that the classifications used are fixed in number and type. If instead we assume that they are themselves random drawings from a possible population of classifications, then the form of the analysis of variance may be altered to reflect this. Such an approach might be useful in attempting to assess the sensitivity of spatial data to the recording units used to measure them. Haggett *et al.* (1977: Chapter 12) illustrate how this approach may be used with data measured on different zonal levels.

When the assumption of fixed levels is removed, it is necessary to modify the error process associated with the model to reflect this. Instead of specifying a Normal process, a gamma process is specified instead. The model thus fitted is a random effects analysis of variance model. This is also termed a components of variance model. For further details, see McCullagh and Nelder (1983).

8.5 CONCLUSIONS

The techniques presented in this chapter have covered a wide range of data analytic problems which are likely to be met in the social and environmental sciences. The techniques associated with linear regression are particularly well known and used in geography, though their formal equivalence with analysis of variance is less well understood. The fact that both techniques can be specified within a single computer package merely by adjusting one or two of the model designation commands provides evidence of this relationship. Similarly, these basic models may be extended to incorporate dummy variables leading either to dummy variable regression or the analysis of covariance.

The requirement that the errors in the two principal models are drawn from a Normal distribution with given mean and constant variance is frequently asserted but is unnecessary. The development of the theory of quasi-likelihood and its formal association with least squares parameter estimation has shown that the key properties of estimators of generalised linear models depend on the constancy of the mean to variance relationship and a lack of autocorrelation. However, though this generalises the quasi-likelihood findings to cover a wider range of models than formerly, it does not remove many of the key difficulties of using these techniques with data which are inherently autocorrelated either in space, time or both. Procedures have been developed to help users of such data operate these models, but to date, these remain reasonably complex. Even so, before one embarks on their use it is important that students and researchers are familiar with the standard uses of the traditional models. GLIM and the generalised family provide a particularly effective way of learning about them.

9

GENERALISED LINEAR MODELS FOR CATEGORICAL DATA

9.1 INTRODUCTION

The concept of a generalised, integrated treatment of linear models may also be applied to the analysis of categorical data. During the last thirty years major changes have taken place in the procedures available for analysing categorical data in the social and environmental sciences. In the 1950s, most of the procedures available tended to be rudimentary and *ad hoc*, a collection of techniques which was difficult to apply or adapt to unusual types of problem. Since then, a number of developments have occurred which have revolutionalised the situation.

First, a number of authors have shown that it is possible to express the information in categorical data sets in the form of linear models, similar in design to those more readily applied to the analysis of continuous data (Bishop *et al.* 1975; Haberman 1974a, 1979; Andersen 1980). (Indeed, this idea has a much longer pedigree, being suggested in papers as early as 1900.) Second, research into the theoretical properties of maximum likelihood estimators during the early 1960s has shown that it is possible to derive such estimators for models which describe information in categorical data tables: contingency tables (Birch 1963). Third, research into generalised linear models in the 1960s and 1970s has shown that regression-type models can be developed specifically for contingency table data and regression models involving categorical response variables, thus linking the traditional concerns of categorical analysis with standard regression analysis (Grizzle *et al.* 1969; Nelder 1974; Freeman 1987).

As a result, three distinct types of linear model have become widely associated with the contemporary approach to the statistical analysis of categorical data:

1 the hierarchical log-linear model,
2 the logit regression model, and
3 the probit regression model.

All three are considered in this chapter. However, before doing so, it is necessary to consider some of the characteristics of contingency tables which need to be accommodated in a linear model if that model is to be of practical value. The main reason for doing this is that contingency table data present many general problems to the researcher. In attempting to solve these, useful advances may be made towards solving the problems associated with categorical regression.

9.2 CONTINGENCY TABLES

9.2.1 Some essential preliminaries

A contingency table is a form of data presentation which is widely associated with social research, and in particular, with questionnaire surveys. The term contingency table applies to data sets which are created by the cross-classification of categorical variables. Thus, given that categorical data may include both nominal and ordinal measurements (see Chapter 3), the following types of data table may be created:

1 fully-nominal contingency tables (Table 9.1);
2 fully-ordinal contingency tables (Table 9.2);
3 mixed nominal–ordinal contingency tables (Table 9.3).

The three data tables presented in Tables 9.1 to 9.3 are created by cross-classifying two categorical variables: car usage and superstore patronage. As all three tables comprise only two categorical variables they are sometimes termed two-way tables. An alternative method of description is to refer to the number of levels or categories associated with each variable. Thus Table 9.1 is a 2×2 table because it has been created by the cross-classification of two binary categorical variables.

The information contained within a contingency table depends entirely on how the cross-classifying variables have been measured and categorised. This is a key issue and reflects the major problems of measurement described in Chapter 3 (section 3.4) concerning the representation of information in low-level measurement classes. For certain types of data, for example sex, there is little ambiguity concerning the number and types of level which should be used to represent the information being considered. However, for most categorical measurements, a degree of subjective judgement is involved. This is manifest in the number, composition and character of the levels which the researchers decide, for whatever reason, are needed in their study. To illustrate this, consider a categorical variable designed to represent types of employment within a country. One possible classification could be created using levels which correspond to broad industrial generalisations of the economy (for example, primary, secondary, tertiary, quaternary), a second could make use of a more

233

detailed, but more numerous, selection of subdivisions. Table 9.4 presents some alternatives based on the standard industrial classification devised for use in the UK from 1980. Table 9.4(a) uses ten industrial 'divisions'; Table 9.4(b) uses thirty-six 'categories' to provide for an even finer discrimination of industries.

Table 9.1 A fully-nominal contingency table

| | | Superstore patronage | |
		No	Yes
	No	66	100
Car usage			
	Yes	54	231

Table 9.2 A fully-ordinal contingency table

| | | Superstore patronage | | |
		Non-user	Light user	Heavy user
	Never	66	71	29
Car usage	Sometimes	29	80	42
	Always	25	61	48

Table 9.3 A mixed nominal–ordinal contingency table

| | | Superstore patronage | | |
		Non-user	Light user	Heavy user
	No	66	71	29
Car usage				
	Yes	54	141	90

It is important to remember that the issue is not just one concerning the number of levels which should be used, it also concerns the quality or character of those levels. In particular, it reflects the meaning to be given to the levels. As a result, the primary problem in the analysis of categorical data is conception not measurement, and it is as well to remember that no amount of technical processing, or refined computer modelling, will rectify errors or ambiguities made at this stage.

9.2.2 Table architecture

To illustrate how the information contained in two binary categorical variables may be organised into a contingency table, consider Table 9.1. This shows the cross-classification of car usage by store patronage. The information on car usage is presented as the rows of the table with the 'No' category corresponding to row 1, and the 'Yes' category to row 2.

Table 9.4 Some alternative industrial classifications (1980 definition)

(a) Division

0 Agriculture, forestry and fishing
1 Energy/water supply industries
2 Extraction/manufacture: minerals/metals
3 Metal goods/vehicle industries, etc.
4 Other manufacturing industries
5 Construction
6 Distribution, hotels/catering; repairs
7 Transport/communication
8 Banking, finance, insurance, leasing, etc.
9 Other services

(b) Categories

01 Agriculture, forestry and fishing
02 Coal extraction
03 Mineral oil and natural gas extraction
04 Mineral oil processing
05 Nuclear fuel production
06 Gas, electricity and water
07 Extraction of other minerals and ores
08 Metal manufacture
09 Manufacture of non-metallic products
10 Chemical industry
11 Production of man-made fibres
12 Manufacture of metal goods
13 Mechanical engineering
14 Manufacture: office machinery/DP equip.
15 Electrical and electronic engineering
16 Manufacture of motor vehicles
17 Shipbuilding and repairing
18 Manufacture: aerospace/transport equip.
19 Instrument engineering
20 Food, drink and tobacco
21 Textiles
22 Leather, footware and clothing
23 Timber and furniture
24 Paper, printing and publishing
25 Other manufacturing
26 Construction
27 Wholesale and distribution
28 Retail distribution
29 Hotels and catering
30 Repair of consumer goods and vehicles
31 Transport
32 Telecommunications
33 Insurance, banking, etc., business services
34 Public administration and defence
35 Medical and other health services
36 Other services – NES

Similarly, the information on superstore patronage is presented as the columns of the table, with the 'No' category corresponding to column 1 and the 'Yes' category to column 2. The cross-classification of two binary variables creates a two-way contingency table containing four cells (i.e., $2 \times 2 = 4$, hence the use of the arithmetic expression to define contingency tables). The numbers associated with these cells represent observed cell frequencies from the researcher's survey. They describe the number of times in a sample of 451 households in Cardiff each of the four combinations of car usage and store patronage occur.

By minor modification, Table 9.1 can be transformed into Table 9.5, by the addition of a third row and column to represent the totals associated with each. These totals are frequently termed marginals in the statistics literature, a term which will be used here. Because the sample size is constant in both Tables 9.1 and 9.5, there are a number of simple relationships which may be exhibited between the observed cell frequencies, the observed row and column marginals, and the observed grand total. These are:

1 The sum of the four observed cell frequencies equals the observed grand total.
2 The sum of the observed row and column frequencies equals the observed row and column marginals.
3 The sum of the observed row or column marginals equals the observed grand total.

Table 9.5 Marginal totals

| | | Superstore patronage | | |
		No	Yes	Total
Car usage	No	66	100	166
	Yes	54	231	285
	Total	120	331	451

These are obvious from the observed table, but it is worth stating them rather pedantically here because models which may be applied to this table must respect and preserve some or all of these observed relationships if they are to be considered adequate. Furthermore, they illustrate that once the grand total and some of the observed marginals and observed cell frequencies are known, any components of the table which are missing can be calculated directly merely by arithmetic.

The latter point is particularly instructive because it means that several different combinations of observed cell frequencies and marginals will satisfy the conditions listed above (Table 9.6). If several solutions exist, then maybe some are more useful than others?

Table 9.6 Some relationships between
observed cells, marginals and the grand total

(a)				Total
		83	83	166
		37	248	285
	Total	120	331	451

(b)				Total
		60	106	166
		60	225	285
	Total	120	331	451

(c)				Total
		0.5	165.5	166
		119.5	165.5	285
	Total	120	331	451

9.2.3 Basic analysis

The fact that the interrelationships between observed cells, marginals and grand total can be satisfied by a variety of different combinations of number suggests that it may be feasible both to describe the patterns between the observed frequencies in some consistent way, and assess rival hypotheses concerning these patterns. The latter point clearly implies that a series of rival hypotheses will generate different patterns of expected cell frequencies.

The traditional approach to the analysis of this type of data is to calculate the chi-square statistic for it. This statistic (previously presented in Chapter 4, section 4.7.3) assesses whether two cross-classifying variables, A and B, may be said to be independent of each other. This term is based on the following probability statements:

$$p(A|B) = p(A)$$

and

$$p(B|A) = p(B)$$

The former states that the probability of A occurring given B is $p(A)$; the latter, that the probability of B occurring given A is $p(B)$. In both statements, the presence of the conditioning variable (B in the first statement and A in the second) does not affect the probability of the other variable. In practical terms this means that knowledge of A (B) does not provide knowledge of B (A). In terms of Table 9.1, this means that information on car usage tells us nothing about store patronage.

In order to assess the hypothesis that car usage and store patronage are

237

indeed independent, a table of cell frequencies which would be expected were they really independent must be generated. Then, the cell-by-cell differences between the expectations under independence and those observed from the survey have to be calculated. The procedure for calculating the expectations is quite simple. For cell 11 it involves multiplying together the marginals for row 1 and column 1 of the observed table and dividing the product by the grand total. That is:

for cell 11: $(166 \times 120) / 451$

For cell 12, the expectation is calculated by multiplying the marginals for row 1 and column 2, and dividing the product by the grand total. The four observed cell frequencies and the expectations under the hypothesis of independence are:

cell	observed frequency	expected frequency	difference
11	66	44.2	21.8
12	100	121.8	−21.8
21	54	75.8	−21.8
22	231	209.2	21.8

The full table of expected cell frequencies under independence is presented in Table 9.7. Notice that the marginals and the grand total have been preserved, that is, they are the same as in the observed table. Notice also that the cell frequencies in the expected table differ from those in the observed table. This provides more evidence to suggest that the patterns of association noticed earlier may be preserved by more than one solution.

Table 9.7 Expected cell frequencies under hypothesis of independence

		Superstore patronage		
		No	Yes	Total
Car usage	No	44.2	121.8	166
	Yes	75.8	209.2	285
	Total	120	331	451

In order to assess whether the cross-classifying variables in the observed table are indeed independent of each other, the differences between the observed cell frequencies and their associated expectations need to be calculated. These are used as input into the chi-square test statistic, an empirically-derived statistic which has the important property that, for given degrees of freedom, it approximates the theoretical chi-square probability distribution (see the discussion in Chapter 5). It is thus possible to use this distribution, whose properties are well known, to provide the test for the independence hypothesis.

The test statistic is given by the formula:

$$X^2 = \text{Sum} \frac{(\text{observed} - \text{expected})^2}{\text{expected}} \qquad (9.1)$$

or

$$X^2 = \Sigma \frac{(o - e)^2}{e}$$

and is calculated as 23.27 for the data under consideration. The number of degrees of freedom associated with this hypothesis is calculated from the expression:

$$df = (\text{rows}-1)(\text{columns}-1)$$

which, for a two-way table, yields a value of 1. The independence hypothesis may now be tested by comparing the calculated value of chi-square with the value expected by chance under the chi-square distribution for 1 degree of freedom. Table 9.8 presents some alternatives associated with different levels of probability. From this, it is possible to see that the value of chi-square calculated from the data exceeds the expected values in the table at the 0.1 significance level. As a value greater than this can only occur once in a thousand by chance, it is reasonable to suggest that the null hypothesis of independence is rejected: car usage and superstore patronage are thus related in some way.

A summary of this analysis using MINITAB is presented in Printout 9.1. The data for the observed table are read into two column variables (C1 and C2) and the statistic calculated using the CHISQUARE command. As output, MINITAB produces expected counts under independence, the chi-square value calculated from the data and its associated degrees of freedom. This should be referred to for comparison when a series of alternative forms of analysis are applied to these data using GLIM.

9.3 LOG-LINEAR RE-ANALYSIS

A comparison of Table 9.6 with Table 9.7 shows that a variety of combinations of observed frequencies will satisfy the interrelationships between the marginals and the grand total. Only one of these corresponds to the hypothesis of independence. This suggests that other hypotheses might also be suggested for the interrelationships in the table.

In the past, it was difficult to test alternative hypotheses to independence in a contingency table in a straightforward and efficient manner. This is no longer the case, as the development of a family of models known as

Table 9.8 Values of the chi-square analysis

Degrees of freedom	Probability values		
	5%	1%	0.1%
1	3.8	6.6	11.0
2	6.0	9.2	14.0
3	7.8	11.3	16.0
4	9.5	13.3	18.0
5	11.0	15.0	21.0
10	18.0	23.0	30.0
15	25.0	31.0	38.0
20	31.0	38.0	43.0
25	38.0	44.0	53.0
30	44.0	51.0	60.0

Printout 9.1 Chi-squared* analysis of Table 9.1

```
£run *minitab
−read c1 c2
−data
      66 100
      54 231
−end
−print c1 c2
  ROW   C1  C2

      1  66 100
      2  54 231

−chisquared c1−c2
```

Expected counts are printed below observed counts

	C1	C2	Total
1	66	100	166
	44.2	121.8	
2	54	231	285
	75.8	209.2	
Total	120	331	451

ChiSq =   10.79+   3.91+
          6.29+   2.28= 23.27

df = 1
−stop

Notes: Commands prefixed by £ are operating system commands
Commands prefixed by − are MINITAB commands

hierarchical log-linear models presents researchers with a much more effective mode of analysis.

9.3.1 Fitting the log-linear model to Table 9.1

A useful way of introducing this family of log-linear models, and illustrating the sort of information they may represent, is to use GLIM to fit a series of alternative models to the observed data in Table 9.1. At this stage it is not necessary to describe the models formally. This will be done in section 9.3.2, after the basic idea of their use has been illustrated. GLIM is used here because MINITAB does not provide facilities for the fitting of log-linear models.

The GLIM commands associated with this analysis are set out in Printout 9.2. In essence, these fall into three general categories of command:

1 commands to facilitate data entry and manipulation ($UNITS to $LOOK);
2 commands to define a model ($YVAR to $ERR); and
3 commands to estimate the selected model and display results from it ($FIT to $STOP).

Printout 9.2 GLIM commands used with Table 9.1

```
$UNITS 4
$DATA N
$READ
66 100 54 231
$FACTOR CAR 2 PAT 2
$CALCULATE CAR=%GL(2,2):PAT=%GL(2,1)
$LOOK N CAR PAT$
$YVAR N
$LINK LOG
$ERR P
$FIT$
$DISPLAY ER$
$FIT CAR+PAT$
$DISPLAY ER$
$FIT +CAR.PAT$
$DISPLAY ER$
$STOP$
```

The $UNIT command is presented first. This is used (as in the examples in Chapter 8) to declare the number of items of data to be handled in the analysis. For contingency table problems the value associated with this command is the number of cells of observed data in the table, hence $UNITS 4. This command also calibrates internal vectors within GLIM to be of length 4. The $DATA command is used next to prepare GLIM for the entry of data. This command creates a variable within the program, termed N, which will be used to store the raw data, that is, the four observed cell frequencies. GLIM does not need to be told that four items will be read into N as this has already been set by the $UNITS command. Having done this, the data are now read into the program from the computer keyboard using the $READ command.

These three commands handle data entry. The next two ($FACTOR and $CALCULATE) handle data manipulation. The former is used to define the presence of the two categorical variables (termed 'factors' in GLIM terminology), CAR and PAT, each of which contains two levels. The latter makes use of the internal system facility (%GL – a facility which generates factor levels within GLIM) to reproduce the architectural structure of the observed contingency table within GLIM. As in the analysis of variance examples in Chapter 8, these are needed because no information on this crucial topic has been read into GLIM as data. As a result, GLIM cannot associate the four data items with the four cells of Table 9.1 unless their relative positions in the table are recreated within the program. The following specifications of %GL are used to recreate the structure of Table 9.1 within GLIM:

1 %GL(2,2), to generate a sequence which runs from 1 to 2 in bundles of 2, namely: 1 1 2 2,
2 %GL(2,1), to generate a sequence which runs from 1 to 2 in bundles of 1, namely: 1 2 1 2.

Each level generated corresponds to a row or column index in the original data table.

The specification of these levels requires researchers to master these commands. To check that everything has been correctly specified, the $LOOK command may be used to generate a printed list of N (the observed cell frequencies) and the factor levels associated with CAR and PAT. The following output should be produced:

N	CAR	PAT
66	1	1
100	1	2
54	2	1
231	2	2

This shows that the observed frequency 66 is allocated to cell 11 (level 1 of CAR and level 1 of PAT), and 100 is allocated to cell 12 (level 1 of CAR and level 2 of PAT). If the factor structure printed at this point is not as perceived in the observed table (for example, if 66 appears to be at cell 12 rather than cell 11), then the factor levels have been generated incorrectly and revisions are required. (Notice that it is possible to combine more than one calculation on the same line, thus avoiding the need for separate CALCULATE commands.)

These commands complete the section on data entry and manipulation. In comparison, the commands needed to specify a particular model are relatively trivial. As suggested in Chapter 7, and illustrated in Chapter 8, three commands are required here:

1 a command to identify the 'response';
2 a command to identify the probability process to be used in inference;
3 a command to link the response to the linear predictor.

The definition of the response in contingency table analysis differs from that encountered in the analysis of continuous data, at least for certain types of contingency table problem. Two separate forms of analysis need to be distinguished here: analyses in which one is interested in 'inter-dependency' in the table, and analyses in which one is interested in 'dependency', that is, in assessing how the variability in one of the categorical factors is conditionally dependent on the variability in the other. The former is the more general analytical problem, frequently termed a symmetric problem, because the pattern of association to be described can vary from row to column and vice versa. The latter is a special-case – an asymmetric problem – in the sense that interest focuses on the pattern of variability in a response variable conditionally dependent on the variability in the explanatory variable (i.e., the 'flow' or direction of association is one way).

For the purposes of this example the table will be treated as symmetric, in other words, ignoring any patterns of dependency which might exist or which might subsequently be highlighted. As neither of the categorising variables is to be treated as the response, it is usual to specify the observed cell frequency (N) instead. This is done using the $YVAR(IABLE) command. The linear predictor for the log-linear models which are to be fitted has already been created using $FACTOR and $CALCULATE. This is linked to the response component in the model using the $LINK command. The most appropriate link for use here is the logarithmic link ($LINK LOG). The final component needed to complete the specification of log-linear models in GLIM is to define a probability process for the estimation of model parameters. The process which is most suitable for log-linear models applied to symmetric tables is the Poisson process. This is specified in GLIM using the $ERROR command, $ERR(OR) P.

Having set up the data in an appropriate form and specified the model structure correctly, the next step is to fit models to the observed data and assess their impact. This is done using the $FIT and $DISPLAY commands. Each $FIT command is followed by a different series of terms, for example, CAR, PAT, CAR+PAT, which correspond to different types of log-linear model. The performance of these is summarised in Printout 9.3,

Table 9.9 Analysis of deviance table for GLIM analysis of Table 9.1

Model	Deviance	df	Change in deviance	Change in df
%GM	157.2	3		
+CAR+PAT	22.72	1	134.48	2
+CAR.PAT	0.0	0	22.72	1

Note: df = degrees of freedom

Printout 9.3 Logged transcription of the GLIM analysis of Table 9.1

```
£r *glim
£Execution begins
  GLIM 3.12 (C)1977 ROYAL STATISTICAL SOCIETY, LONDON

  $UNITS 4
  $DATA N
  $READ
  66 100 54 231
  $FACTOR CAR 2 PAT 2
  $CALCULATE CAR=%GL(2,2):PAT=%GL(2,1)
  $LOOK N CAR PAT$
    1    66.00    1.000    1.000
    2    100.0    1.000    2.000
    3    54.00    2.000    1.000
    4    231.0    2.000    2.000
  $YVAR N$LINK LOG$ERROR P$
  $FIT$
            SCALED
   CYCLE  DEVIANCE  DF
      4    157.2     3

  $D E R$
          ESTIMATE        S.E.      PARAMETER
    1       4.725    0.4709E−01  %GM
    SCALE PARAMETER TAKEN AS 1.000

    UNIT  OBSERVED  FITTED    RESIDUAL
      1       66     112.7     −4.403
      2      100     112.7     −1.201
      3       54     112.7     −5.533
      4      231     112.7      11.14

  $FIT CAR+PAT$
            SCALED
   CYCLE  DEVIANCE  DF
      4    22.72     1

  $D E R$
          ESTIMATE     S.E.       PARAMETER
    1       3.788    0.1102       %GM
    2       0.5405   0.9764E−01   CAR(2)
    3       1.015    0.1066       PAT(2)
```

244

SCALE PARAMETER TAKEN AS   1.000

UNIT	OBSERVED	FITTED	RESIDUAL
1	66	44.17	3.285
2	100	121.8	−1.978
3	54	75.83	−2.507
4	231	209.2	1.510

$LOOK %X2$
1	23.27

$FIT +CAR.PAT$

		SCALED	
CYCLE		DEVIANCE	DF
2		0.4667E−09	0

$D E R$

	ESTIMATE	S.E.	PARAMETER
1	4.190	0.1231	%GM
2	−0.2007	0.1835	CAR(2)
3	0.4155	0.1586	PAT(2)
4	1.038	0.2191	CAR(2).PAT(2)

SCALE PARAMETER TAKEN AS   1.000

UNIT	OBSERVED	FITTED	RESIDUAL
1	66	66.00	0.7748E−05
2	100	100.0	0.9537E−05
3	54	54.00	0.0
4	231	231.0	0.1449E−04

$STOP$

Note: £ prefixes are operating system commands
   $ prefixes are GLIM commands

which is a transcript of the GLIM computing session, and in Table 9.9, which displays some of the output as a summary analysis of deviance table. As before, the term deviance is used to refer to variability within a data set which has not been accounted for by the model, that is, residual variability.

The first of the $FIT commands fits a single parameter effect to the summary log-linear model, a parameter effect which represents the grand mean of the observed data. This is the simplest form of log-linear model which can be specified. It calculates a value for each of the cells in the table of expected frequencies from the expression:

expected cell frequencies   =   average of the observed total

The table of expected cell frequencies produced as a result of this model is displayed in Table 9.10. Notice the grand total is identical to that in the observed table but that the row and column marginals are not the same. This has happened because no account has been taken of the marginal distributions of the observed table. As this is the simplest log-linear model available, its explanatory power is very low, not surprisingly given the fact

Table 9.10 Expected frequencies under null model

			Total
	112.75	112.75	225.5
	112.75	112.75	225.5
Total	225.5	225.5	451

Note: This model preserves the observed grand total

that the expected frequencies merely reflect the sample size and the number of classifying cells. The amount of unexplained variability left after this has been fitted is listed as the first entry under the deviance heading (157.2 for 3 degrees of freedom).

This figure is the log-linear model equivalent of the total sum of squares measure which is associated with the linear regression and analysis of variance models. It indicates the maximum amount of variability which needs to be accounted for by parameters in the model. (It is also frequently referred to as the maximum log-likelihood statistic, or as G^2, two alternative presentations referring to the same idea.) The column headed degrees of freedom indicates the number of independent items of information associated with the fitted model. Thus for the addition of the grand mean effect, a maximum value of 157.2 has been calculated for the unexplained variability in the data which is associated with three degrees of freedom.

An improvement in the explanatory power of the fitted model may be made by adding further parameters. This is done using the second $FIT command ($FIT CAR + PAT). This model fits what are termed the main effects of the two cross-classifying factors on the cell frequencies. The expected cell frequencies under this model are presented in Table 9.11. Notice, this time, that the marginals and the grand total of the expected table have reproduced their counterparts in the observed table, and that the expected cell frequencies are identical to those in Table 9.7, the table of expected cell frequencies produced by the CHISQUARE command in MINITAB, that is, under the hypothesis of independence. This model is therefore assessing the same type of hypothesis as is tested using the chi-square statistic.

Table 9.9 shows that under independence the deviance measure has been reduced from 157.2 to 22.72 for the addition of two new parameters.

Table 9.11 Expected frequencies under independence model

			Total
	44.17	121.83	166
	75.83	209.17	285
Total	120	331	451

Note: This model preserves the observed grand total and row and column marginals

As a result there is now only one degree of freedom remaining in the table. Notice that the deviance measure associated with the model is very similar to that produced for chi-square in MINITAB. By typing the command, $LOOK %X2, we can tell GLIM to generate the chi-square statistic associated with this log-linear model (%X2 is a system vector within GLIM). This shows that the value is identical to that produced by MINITAB.

The final log-linear model fitted to the data in Table 9.1 adds what is termed a two-way interaction effect (+ CAR . PAT) to the independence model. This is used to describe the fact that the classification of any specific row (or column) of the observed data may depend to some extent on their column (or row) classification. The effect of this new model on the expected cell frequencies is given in Table 9.12. Notice that this table is identical in every respect to the observed data table. Similarly, the effect of this on deviance is to reduce the measure to zero for zero degrees of freedom, indicating that there is now no residual variability within the table left to be explained. What in fact has happened is that a parameter has been fitted to represent every possible form of interdependency in the observed data. The model therefore represents the observed data exactly. Because of this, this saturated model corresponds to the most complex form of log-linear model which may be fitted to the table.

Table 9.12 Expected frequencies under saturation

			Total
	66	100	166
	54	231	285
Total	120	331	451

Note: All table entries correspond to those in table of observed frequencies

9.3.2 Some notation

The grand mean effects model and saturated model represent the lower and upper ends of a range of possible log-linear models which may be fitted to a contingency table. In the former case, only one parameter effect is used to calculate the expected cell frequencies; in the latter case, every parameter effect which could be used is. Neither summarises the observed data in a parsimonious form, that is, in as few parameters as possible. A more effective description may exist by selecting a model which contains more parameters than the grand mean effects model, but fewer than the saturated model. In the two-way table independence is one possibility, but there are other possibilities, for example, models which include only one of the two main effects at any one time.

In order to describe the full range of models some notation is needed.

The following are three possible notation schemes (Table 9.13). The first is based on the notation scheme devised by Wilkinson and Rogers (1973) and which is widely used in the GLIM documentation. The second makes use of a scheme devised by Birch (1963) and popularised, in an amended form, by Bishop *et al.* (1975). The third makes use of the scheme based on Goodman (1970). It is of little consequence which of these is used as they have been designed to represent the same types of parametric effects in models. The full range of models available for a two-way table is summarised in Table 9.14, where model 1 is the grand mean effects model, 4 is independence, and 5 is saturation. Models 2 and 3 correspond to the non-comprehensive main effects models in which only one main effect is included at one time. The effects of these on the expected cell frequencies are presented in Table 9.15. Notice that, in these, the grand total and one set of marginals are preserved.

Table 9.13 Notation schemes for log-linear models

Parameter effect	GLIM	Birch	Goodman
Grand mean	%GM (1)	U	λ
Main effect of			
CAR	CAR	$U_{1(i)}$	λ_c^C
PAT	PAT	$U_{2(j)}$	λ_p^P
Interaction between			
PAT and CAR	CAR*PAT	$U_{12(ij)}$	λ_{cp}^{CP}

Note: The notation CAR*PAT is shorthand for %GM+CAR+PAT+CAR.PAT. Similarly, $U_{12(ij)}$ implies the presence of $U + U_{1(i)} + U_{2(j)}$ and λ_{cp}^{CP}, the presence of $\lambda_1 \lambda_c^C, \lambda_p^P$ In GLIM 3.77, the %GM term is replaced by the number 1 in printed output

Table 9.14 Log-linear models for a two-way contingency table

Model	GLIM	Birch	Goodman
1	%GM	U	λ
2	%GM + CAR	$U + U_{1(i)}$	$\lambda + \lambda_c^C$
3	%GM + PAT	$U + U_{2(j)}$	$\lambda + \lambda_p^P$
4	%GM + CAR + PAT	$U + U_{1(i)} + U_{2(j)}$	$\lambda + \lambda_c^C + \lambda_p^P$
5	%GM CAR*PAT	$U + U_{1(i)} + U_{2(j)}$ $+ U_{12(ij)}$	$\lambda + \lambda_c^C + \lambda_p^P$ $+ \lambda_{cp}^{CP}$

Note: The GLIM notation CAR*PAT corresponds to:

%GM + CAR + PAT + CAR.PAT

All the notation schemes presented in Table 9.13 develop from a core expression which is set out in model 1 of Table 9.14, the grand mean effects model. This development indicates that the log-linear models to be considered here are hierarchical, that is, the higher-order effects and

Table 9.15 Expected frequencies under non-comprehensive main effects models

(a) *$FIT CAR$*

			Total
	83	83	166
	142.5	142.5	285
Total	225.5	225.5	451

Note: This model preserves the observed row marginals

(b) *$FIT PAT$*

			Total
	60	165.5	225.5
	60	165.5	225.5
Total	120	331	451

Note: This model preserves the observed column marginals

interactions can only be fitted when their lower-order relatives are also present in the model. For this reason, it is not possible to include an interaction term in a non-comprehensive main effects model; the interaction can only be fitted when the main effects from which it is based are also in the model. This hierarchical arrangement does not occur by chance but is a direct result of the definition of each set of effects.

9.3.3 Defining and interpreting parameter effects

Log-linear models illustrate what are termed over-parameterised models. This means that they contain more parameters than independent pieces of information to estimate them. Consequently, in order to define the parameter effects, that is the grand mean, main effects and interactions, it is necessary to apply some system of constraints to the models (see Searle 1971 for details of the general issue). These affect the definition of the parameter effects which may be applied.

Two systems of constraints are frequently used with log-linear models. These are:

1 corner-weighted constraints; and
2 centre-weighted constraints.

The former are used in GLIM to calculate estimates of any over-parameterised model, not just log-linear models. The latter, sometimes known as the usual constraints (Payne 1977), are to be found in computer programs such as ECTA (Fay and Goodman 1975) and BMDP. However, they are also implicitly assumed to be used by many of the authors of standard textbooks on contingency table analysis (for example, Bishop *et*

249

al. 1975). Corner-weighted constraints work by setting one parameter in each parameter effect to zero and defining the remaining parameters as contrasts with it. Centre-weighted constraints, on the other hand, assume that the sum of the parameters in each parameter effect is zero. Each parameter is thus defined as a contrast with the average for the parameter effect as a whole.

Given this distinction, the following definitions may be made for the three types of parameter effect introduced previously:

(a) *corner-weighting*

1 *grand mean*: the expected frequency in cell 11 of the contingency table;
2 *main effects*: the difference in the expected cell frequencies of being at level 2 of CAR rather than level 1, or at level 2 of PAT rather than level 1;
3 *interactions*: the difference in the expected cell frequencies of being simultaneously at level 2 of CAR and PAT (that is, cell 22) rather than level 1.

(b) *centre-weighting*

1 *grand mean*: the average expected cell frequency in the contingency table;
2 *main effects*: the difference in the expected cell frequencies of being at level *i* of CAR, or *j* of PAT, rather than the average;
3 *interactions*: the difference in the expected cell frequencies of being simultaneously at level *i* of CAR and *j* of PAT rather than the average.

These differences in definition reflect the constraints used. For the models specified in GLIM the following constraints apply:

main effects: $\lambda_1^{CAR} = \lambda_1^{PAT} = 0$

interactions: $\lambda_{1j}^{CARPAT} = \lambda_{i1}^{CARPAT} = 0$

(note that the use of the first parameter as the bench-mark is purely arbitrary), whereas using the centre-weighting, the following are used instead:

main effects: $\Sigma \lambda_i^{CAR} = 0 \quad \Sigma \lambda_j^{PAT} = 0$

interactions: $\underset{i}{\Sigma} \lambda_{ij}^{CARPAT} = \underset{j}{\Sigma} \lambda_{ij}^{CARPAT} = 0$

The constraints themselves are a necessary evil. Their presence does not affect the calculation of the expected frequencies or the deviance measures under any of the hypotheses listed previously, but the estimates of the individual parameters are affected (Holt 1979). It follows therefore that considerable attention should be paid when using parameter estimates, though readers should note that few authors advocate their use, and Holt (1979) urges strongly against it.

9.3.4 Deviance and degrees of freedom

Having fitted a series of log-linear models it is important to determine which represents the observed data most effectively. This involves assessing the overall performance of the model against a theoretical norm. The procedure is very similar to that used with the chi-square test in section 9.2, in that a measure of goodness-of-fit is compared against a value expected by chance for identical degrees of freedom. The observed effect of a model is determined by comparing the change in the deviance measure as parameter effects are added or removed. Similarly, as the number of effects changes, so too does the number of individual parameters which are fitted. Thus the deviance measures generated by GLIM for specific log-linear models reflect the number of degrees of freedom remaining in the data.

As noted in Chapter 7 the deviance measure for given degrees of freedom is approximately distributed as chi-square. It is thus possible to obtain expected values of chi-square for given degrees of freedom which may be compared with those observed. Values observed which exceed those expected indicate that the model being fitted does not correspond to the data and that its associated null hypothesis is inappropriate. For Table 9.1, these tests of overall fit suggest the following (Table 9.16):

1 rejection of the null hypotheses associated with the grand mean, non-comprehensive main effects and independence models;
2 acceptance of the saturated model.

Though the degrees of freedom are produced as output, it is important to note how they are calculated. For a 2×2 table there is a maximum of 4 degrees of freedom. In fitting the grand mean effects model, a single parameter is fitted to the table, thus reducing the degrees of freedom by 1. Similarly, in fitting each main effect, a single parameter is fitted (the first parameter in each effect being constrained), each reducing the degrees of freedom by 1. (The independence model in which both main effects and

Table 9.16 Critical values for Table 9.1

Model	Change in deviance	Change in df	Critical values		
			5%	*1%*	*0.1%*
% GM + CAR + PAT	134.48	2	6.0	9.2	14
% GM + CAR + PAT + CAR.PAT	22.72	1	3.8	6.6	11

Table 9.17 Calculation of degrees of freedom for two-way contingency tables

Model	Parameter effects added	Degrees of freedom
Grand mean	%GM	$IJ-1 = 3$
Non-comprehensive	+CAR	$IJ - \{1+(I-1)\} = 2$
main effects	or	
	+PAT	$IJ - \{1+(J-1)\} = 2$
Independence	+CAR+PAT	$IJ - \{1+(I-1)+(J-1)\} = 1$
Saturated	+CAR.PAT	$IJ - \{1+(I-1)+(J-1)+(I-1)(J-1)\} = 0$

Note: IJ is the product of the number of rows and columns. Degrees of freedom presented are for a 2×2 table

grand mean are present thus has 1 degree of freedom.) Finally, in fitting the single interaction term the last degree of freedom of the table is used up, resulting in a model with zero degrees of freedom (Table 9.17).

9.3.5 Beyond summary measures

Knowledge from the summary statistics that CAR and PAT are not independent is useful but further information may be obtained by looking at the odds of store patronage at both levels of car usage. For non-car users, the odds of patronage against non-patronage are:

odds1 = 100/66

Similarly, for car users, the odds of patronage against non-patronage are:

odds2 = 231/54

The cross-product ratio – the odds of patronage against non-patronage for both levels of car usage – is defined as the ratio of odds1 to odds2:

CPR = odds2/odds1
    = (231/54) / (100/66)
    = (231×66) / (54×100) = 2.83

This value states that the odds of using a superstore are 2.83 times higher for car users than for non-car users, reflecting the mobility differences between the two groups. This ratio of odds may also be presented as a ratio of log-odds. The logarithmic cross-product ratio for Table 9.1 is 1.04.

Because the data in the table are obtained from a sample, it is useful to obtain confidence intervals for the odds and log-odds as an indication of the significance of the difference between car users and non-users. To calculate these we first need to calculate the estimated sample standard error for both measures. For the log-odds ratio this is given as (after Fleiss 1981):

252

$$\sqrt{1/66 + 1/100 + 1/54 + 1/231} = 0.219$$

yielding a $100(1-.05)$ per cent confidence interval estimate of:

$$1.04 + 1.96(0.219)$$

The limits of this 95 per cent confidence interval correspond to 0.1815 and 1.898. For the odds ratio, the corresponding 95 per cent confidence intervals can be shown to be 1.2 and 6.67. As independence between CAR and PAT corresponds to an odds ratio value of 1 (or log-odds of 0), values which do not lie in either 95 per cent confidence interval, we can be even surer that the two variables are not independent.

The use of the odds and log-odds ratios as an adjunct to log-linear models is a valuable check on the performance of the models. Both ratios are useful measures of cell-by-cell variability within a contingency table and possess a number of attractive properties:

1 they are invariant to the multiplication of row or column values by a constant;
2 they yield identical information if the orientation of the table is changed, that is if the rows and columns are swapped, or if the order of the cells in the rows or columns changes.

It is also possible to rewrite them both in terms of log-linear models (see Bishop *et al.* (1975) and Payne (1977) for further details).

The concept of odds ratios may be extended to accommodate interaction effects in contingency tables which are more complex than the two-way. In multi-way tables a third or may be even a fourth dimension is added to the design, significantly increasing the number and type of interactions present in the table. If these are to be modelled effectively, a consistent definition of higher-order interaction needs to be created.

The simplest extension of the simple two-way table is the $2 \times 2 \times 2$ table, created by the cross-classification of three binary variables. Bartlett (1935) showed that the hypothesis of no second-order interaction (that is, a hypothesis which tests for the presence of an interaction between the three variables) could be described by the following cross-product ratios:

$$\frac{F_{111}\,F_{221}}{F_{121}\,F_{211}} = \frac{F_{112}\,F_{222}}{F_{122}\,F_{212}} \tag{9.2}$$

where F_{ijk} refers to the expected cell frequencies at levels i of the row variable, j of the column variable, and k of the third variable. If a three-way interaction does exist, then the two cross-product ratios will not be equal. Roy and Kastenbaum (1956) and Birch (1963) developed this idea to accommodate three-way interaction effects in general $I \times J \times K$ tables. For these, the Bartlett formula is rewritten as:

$$\frac{F_{rst}\ F_{ijt}}{F_{ist}\ F_{rjt}} = \frac{F_{rsk}\ F_{ijk}}{F_{isk}\ F_{rjk}} \tag{9.3}$$

where

$$1 < i < r - 1$$
$$1 < j < s - 1$$
$$1 < k < t - 1$$

The main attraction of this approach is that it may be rewritten in the form of a linear-in-parameters model which is additive in the natural logarithmic scale:

$$\begin{aligned}\ln F_{ijk} = {}&\ln F_{rjk} + \ln F_{isk} + \ln F_{ijt} - \ln F_{ist} - \ln F_{rjt}\\ &- \ln F_{rsk} + \ln F_{rst}\end{aligned} \tag{9.4}$$

in other words, a log-linear model. Birch also showed that it was possible to produce maximum likelihood parameter estimates for this form of model, and to develop it further, so that it might represent higher-order interactions in more complex contingency tables. (For a discussion of some contemporary studies into the nature of interaction in higher-order designs, see, among others, Darroch 1974, Whittemore 1978, Snee 1982, and Cox 1984.)

9.4 MULTI-WAY CONTINGENCY TABLES

9.4.1 Models for the three-way design

In comparison with the chi-square statistic or other traditional measures of association, the log-linear family of models is a particularly flexible and instructive technique. The value of log-linear modelling is even more ably demonstrated when the table to be described is multi-way in design rather

Table 9.18 A multi-way contingency table

		TYPE					
		S ENV			M ENV		
		D	R	N	D	R	N
	Far	28	1	14	20	5	16
LOCATION	Medium	97	9	33	41	6	27
	Near	75	8	37	22	1	19

Source: Based on Table 1 of Brodsky and Hakkert (1985)

Notes: TYPE: S = Single vehicle/pedestrian accidents
   M = Multiple vehicle accidents

ENV(IRONMENT): D = Daytime accidents
   R = Daytime accidents on rainy days
   N = Night-time accidents
LOCATION: Distance to nearest hospital

than two-way. To illustrate this, consider the data in Table 9.18 which were collected as part of a study of paramedical services in Haifa, Israel (Brodsky and Hakkert 1985).

The authors were particularly interested in the factors which led bystanders to summon emergency ambulance assistance to the scenes of road traffic accidents. Three variables were isolated from the log-books of the ambulance services and from police records:

1 T – a binary variable representing accident type, and distinguishing between accidents involving single vehicles and/or pedestrians, and multiple vehicle accidents.
2 L – a three-way variable representing the location of the accident with respect to the nearest hospital. The levels used in classification were 'Near', 'Medium distance', and 'Far'.
3 E – a three-way variable representing the environmental conditions associated with the accident. The levels used correspond to accidents occurring during the day, during days on which it is raining, and during the night.

In terms of the log-linear framework the following seven types of models may be fitted to this table:

1 saturation,
2 pairwise association,
3 conditional independence,
4 multiple independence,
5 mutual independence,
6 non-comprehensive models, and
7 null-effects (grand mean) model.

These are written out algebraically in Table 9.19. (Readers should note that effects such as LE and EL or LET and TEL are formally identical and achieve the same influence on deviance.) Models 1, 5, 7 and some of type 6 are common to the analysis of the two-way design. The others are, however, specific to testing interaction effects between two-variables given the presence of the third. Table 9.20 illustrates the effects of some of these models on the calculation of the expected cell frequencies, and summarises the hypotheses which are being tested by them.

Though there are only seven model types, there are actually nineteen distinct models which may be fitted to the table. The reason for this difference is that the number of ways of forming two-way interactions increases rather more than proportionally to the increase in the number of dimensions. The difference widens even more greatly if a fourth or fifth dimension is added.

Table 9.19 Log-linear models for the three-way design

Model type	Terms included	Terms excluded
1 Saturation	%GM+L+T+E+LT+LE+TE+LET	
2 Pairwise association	%GM+L+T+E+LT+LE+TE	LET
3 Conditional	%GM+L+T+E+LT+LE	TE+LET
independence	%GM+L+T+E+LT+TE	LE+LET
	%GM+L+T+E+TE+LE	LT+LET
4 Multiple	%GM+L+T+E+LT	LE+LT+LET
independence	%GM+L+T+E+TE	LT+LE+LET
	%GM+L+T+E+LE	LT+TE+LET
5 Mutual independence	%GM+L+T+E	LT+LE+TL+LET
6 Non-comprehensive	%GM+L+T+LT	E+LE+ET+LET
models	%GM+L+E+LE	T+LT+ET+LET
	%GM+E+T+TE	L+LT+LE+LET
	%GM+L+T	E+LE+LT+TE+LET
	%GM+L+E	T+LE+LT+TE+LET
	%GM+E+T	L+LE+LT+TE+LET
	%GM+L	E+T+LE+LT+TE+LET
	%GM+T	L+E+LE+LT+TE+LET
	%GM+E	L+T+LE+LT+TE+LET
7 null-effects model	%GM	L+E+T+LE+LT+TE+LET

Note: L = main effect of location
E = main effect of environment
T = main effect of accident type
LE, LT, TE = two-way interactions
LET = three-way interaction

Source: After O'Brien (1989)

Table 9.20 Expected cell frequencies under different log-linear models

(a) Pairwise association model (%GM+L+T+E+LT+LE+TE)

		TYPE					
		S ENV			M ENV		
		D	R	N	D	R	N
	Far	27.5	2.8	12.8	20.5	3.3	17.2
LOCATION	Medium	96.4	8.9	33.7	41.7	6.1	26.3
	Near	76.2	6.3	37.5	20.8	2.7	18.5

Note: Each pair of variables is related, but is unaffected by the presence of the third variable

(b) Conditional independence model (%GM+L+T+E+LT+TE)

		TYPE					
		S ENV			M ENV		
		D	R	N	D	R	N
	Far	28.5	2.6	12.0	21.7	3.1	16.2
LOCATION	Medium	92.1	8.3	38.7	39.1	5.7	29.2
	Near	79.5	7.2	33.4	22.2	3.2	16.6

Note: A pair of variables are independent given the presence of the third variable. In this case, L and E given T

(c) Multiple independence model (%GM+L+T+E+LT)

		TYPE					
		S ENV			M ENV		
		D	R	N	D	R	N
	Far	26.5	2.8	13.7	25.3	2.7	13.1
LOCATION	Medium	85.7	9.1	44.2	45.6	4.8	23.5
	Near	74.0	7.8	38.2	25.9	2.8	13.4

Note: Two variables included as a joint variable are independent of the third

(d) Mutual independence model (%GM+L+T+E)

		TYPE					
		S ENV			M ENV		
		D	R	N	D	R	N
	Far	34.1	3.6	17.6	17.7	1.9	9.1
LOCATION	Medium	86.4	9.1	44.6	44.9	4.8	23.2
	Near	65.7	7.0	33.9	34.2	3.6	17.6

Note: All three variables are independent

(e) Non-comprehensive model (%GM+L+T+LT)

		TYPE					
		S ENV			M ENV		
		D	R	N	D	R	N
	Far	14.3	14.3	14.3	13.7	13.7	13.7
LOCATION	Medium	46.3	46.3	46.3	24.7	24.7	24.7
	Near	40.0	40.0	40.0	14.0	14.0	14.0

Note: All environmental categories are equally likely given location and accident type

(f) Non-comprehensive model (%GM+L+T)

		TYPE					
		S ENV			M ENV		
		D	R	N	D	R	N
	Far	18.4	18.4	18.4	9.6	9.6	9.6
LOCATION	Medium	46.7	46.7	46.7	24.3	24.3	24.3
	Near	35.5	35.5	35.5	18.5	18.5	18.5

Note: All categories of environment are equally likely given location and accident type. Location and accident type are independent

(g) Non-comprehensive model (%GM+T)

		TYPE					
		S ENV			M ENV		
		D	R	N	D	R	N
	Far	33.6	33.6	33.6	17.4	17.4	17.4
LOCATION	Medium	33.6	33.6	33.6	17.4	17.4	17.4
	Near	33.6	33.6	33.6	17.4	17.4	17.4

Note: Combinations of environment and accident type are equally likely given location

(h) Null-effects model (%GM)

		TYPE					
		S ENV			M ENV		
		D	R	N	D	R	N
	Far	25.5	25.5	25.5	25.5	25.5	25.5
LOCATION	Medium	25.5	25.5	25.5	25.5	25.5	25.5
	Near	25.5	25.5	25.5	25.5	25.5	25.5

Note: All combinations of location, accident type and environmental type are equally likely

9.4.2 Finding a parsimonious model

Of the nineteen different models which may be fitted to Table 9.18, several may represent its general features perfectly adequately. As a result any of these could be used to describe the patterns in the table. However, this is rather crude and arbitrary. A more sensitive approach is to find the most parsimonious model for the table, that is, the model which represents the key features of the table in as few parameters as possible.

In order to find this parsimonious model researchers are obliged to fit and compare a considerable number of alternative model forms. This can be time-consuming and expensive. Consequently, a number of selection strategies have been suggested which attempt to make the process more efficient, but which also identify the final model accurately. Most of these procedures are based on some form of stepwise process, involving either the addition of parameter effects to a simple model such as independence, or their deletion from a complex model such as saturation. However, any of the alternatives listed in Table 9.19 may be chosen as the base model instead of these, so long as there are reasonable grounds for justifying it.

There are many different types of stepwise procedure which may be used. Two of the most useful techniques are Aitkin's simultaneous testing procedure (STP), and Brown's screening strategy. Both of these are relatively easy to understand and to implement using popular commercial computer packages. For example, the STP is particularly useful if the analysis has taken place using GLIM. Conversely, Brown's approach has been incorporated as an option within the P4F module of BMDP (Dixon 1983), and can easily be extended for use with ECTA (Fay and Goodman 1975). Wrigley (1985) discusses the merits and demerits of a number of other procedures which also exist.

9.4.3 Simultaneous testing of interactions using STP

The approach developed by Murray Aitkin (1978, 1979, 1980) is based upon the simultaneous testing of families of terms, for example, all two-way interactions. It incorporates the following four steps:

1 The calculation of a Type 1 error rate for the hypothesis that all the interactions applicable to the table are insignificant.
2 The calculation of Type 1 error rates for the hypothesis that specific families of interactions are insignificant.
3 The calculation of expected levels of deviance reduction associated with specific families.
4 The elimination of families whose observed effect on deviance reduction is less than that expected at stage 3.

The idea behind this form of group testing is that, in general, families of effects such as two-way or three-way interactions are more likely to contribute to the reduction of deviance than single effects. Therefore, it is more efficient to test the family prior to testing specific effects.

The overall influence of each of the families appropriate to Table 9.18 is summarised as an analysis of deviance table (Table 9.21). Each line of this deviance table describes how the value for deviance changes with the addition of specific families of effects. As more families are added, the value for deviance decreases until, with saturation, it equals 0. At the same time, the value for the degrees of freedom also decreases. It too equals 0 when the saturated model is fitted.

The table of deviance represents the observed outcome of adding specific terms to the previous log-linear model. Thus adding the main effects

Table 9.21 Analysis of deviance table for log-linear analysis of Table 9.18

Model	Deviance	DF	Change Deviance	DF
%GM	369.80	17		
%GM+L+T+E	27.84	12	341.96	5
%GM+L+T+E+ LT+LE+ET	4.53	4	23.31	8
%GM+L+T+E+ LT+LE+ET+ LET	0.0	0	4.53	4

259

to the grand mean effects model reduces deviance by 341.96 but requires the addition of five extra parameters to the grand mean model. Similarly, the addition of the two-way effects to the model containing grand mean and main effects reduces deviance by 23.31 for the addition of eight extra parameters. Finally, the addition of the three-way effects to the model containing grand mean, main effects and two-way interactions reduces deviance by 4.53 for the addition of four extra parameters. How many, if any, of these families of effects are actually reducing deviance significantly?

In order to assess this, the observed effects on deviance reduction need to be compared with amounts expected by chance from families of identical size to those actually fitted to the table. These expected values are calculated from tables of the chi-square distribution for given degrees of freedom. However, unlike the previous use of these tables in section 9.2.3 to test the hypothesis of independence, the level of significance used in testing is not arbitrary, but reflects the global Type 1 error rate considered suitable for the table. For the hypothesis that all interactions are insignificant, this error rate may be calculated from the formula:

$$y = 1 - (1 - a)^{2^r - r - 1} \qquad (9.5)$$

where

  a represents a significance level (e.g., 5%, 10%), and
  r represents the number of dimensions in the table (in this case three, because it is created by the cross-classification of three categorical variables)

In effect, this error rate is used to test the null hypothesis that the three two-way interactions – location and accident type (LT), location and environmental type (LE), and environmental type and accident type (ET) – and the single three-way interaction (LET), are all insignificant. For a = 0.05 the Type 1 error rate is calculated as 0.185; for a = 0.1, it is 0.34. Aitkin advises that a value between 0.25 and 0.5 will generally be sufficiently sensitive for this global error. Given this, the value of a = 0.1 will be used here.

Having established the global error rate, error rates for each family of effects may now be calculated, beginning with the three-way effects. The appropriate rate for this family is given from the formula:

$$y_3 = 1 - (0.9)^1 = 0.1 \qquad (9.6)$$

where the superscript corresponds to the single three-way effect which is being tested. At the global error rate suggested previously, a family error rate of 0.1 is established for the three-way effect. This is used in conjunction with tables of the chi-square distribution to estimate an expected level of deviance reduction of 7.78 for the three-way family. Algebraically, the chi-square measure used is:

$$\chi^2_{4,\,0.1} \tag{9.7}$$

where the first subscript refers to the degrees of freedom associated with the family, and the second to the family error rate. As the expected value is greater than that observed for this family, we can assume that the three-way effect is insignificant at this error rate.

Having tested and eliminated the three-way effect, the next step in the strategy is to test the two-way effects. The same general procedure is adopted, except that the two-way and three-way families are pooled together to form a combined family which is tested as a whole. Aitkin's reason for this modification is that it provides a stronger test than one applied solely to the two-way effects without increasing its overall size. The error rate for the combined family is:

$$y_{2,3} = 1 - (0.9)^4 = 0.34 \tag{9.8}$$

which yields an expected level of deviance reduction of 14.2, calculated from the chi-square distribution at the 0.34 error level for twelve degrees of freedom. By comparing this with the observed level of deviance reduction of 23.31, it is immediately clear that not all of the two-way effects may be eliminated from the final model.

Having found a family of effects which appears to be significant, there is nothing further to be gained by testing lower-order families for significance en bloc. If every two-way effect is to be retained in the final log-linear model, it follows from the hierarchy principle that all lower-order relatives must also be present. However, the fact that the two-way family has been shown to be significant does not mean that every two-way interaction in the family is significant. Indeed, it may be that the principal source of the family's influence rests with only one or two interactions. If this can be shown to be the case, then it may be possible to eliminate some of these effects as well as the three-way interaction.

An effective way of assessing the significance of the two-way effects is to refit them to the main effects model individually. This allows us to produce a more detailed analysis of deviance table (Table 9.22). Effects may be eliminated from this table from the bottom up until the critical level of deviance reduction is exceeded (the critical level being the level of deviance expected by chance from the pooled test). This suggests that the three-way effect and the two-way effect between accident type and environment should be excluded, leaving the following log-linear model:

$$\%GM+L+T+E+LT+LE \tag{9.9}$$
(deviance of 12.31 for six degrees of freedom)

One of the main reasons for assessing parameter effects in families rather than individually is to minimise so-called order-of-entry effects. These

Table 9.22 Revised analysis of deviance table for log-
linear analysis of Table 9.18

Model	Deviance	DF	*Change* Deviance	DF
%GM	369.80	17		
%GM+L+T+E	27.84	12	341.96	5
Main effects				
+LT	15.08	10	12.76	2
+LE	12.31	6	2.77	4
+ET	4.53	4	7.78	2
Eliminated terms	0.0	0	4.53	4

reflect marginal and conditional relationships between terms which may exert an influence on deviance reduction over and above that unique to each effect. The main implication here is that the observed effect on deviance associated with a specific term may depend, to some extent, on its relationships with terms which have already been included in the model. As a result, the order in which terms are added to the base model may ultimately determine which terms are eliminated. Clearly, this is rather crude.

A better procedure to adopt is as follows. As an initial step, fit the two-way effects in any order and calculate their standardised regression coefficients (srcs). These are the ratios of the parameter estimates to their estimated standard errors, and are produced by GLIM in response to the $DISPLAY E command. The terms may then be refitted to the main effects model in src order, beginning with the largest first. This leads to the production of a third analysis of deviance table (Table 9.23) in which the most important terms on the basis of their srcs are fitted first, and the least important fitted last. Table 9.23 shows that the three-way interaction and two-way interaction between location and environment may be eliminated, producing a model with deviance of 7.29 for eight degrees of freedom. This is an improvement on the previous model.

Table 9.23 Analysis of deviance table for log-linear
analysis of Table 9.18 (second revision)

Model	Deviance	DF	*Change* Deviance	DF
%GM	369.80	17		
%GM+L+T+E	27.84	12	341.96	5
Main effects				
+LT	15.08	10	12.76	2
+ET	7.29	8	7.79	2
+LE	4.53	4	2.76	4
Eliminated terms	0.0	0	4.53	4

Ideally, this revised order of entry should result in deviance decreasing by decreasing amounts, indicating that the latter terms contribute less to deviance reduction than the former terms. Should this not be observed, some further refitting may be needed until the order is sufficiently stable. Once an acceptable order has been established, elimination may begin from the bottom up as before. As the order in Table 9.23 is stable, the best-fitting model for Table 9.18 thus appears to be a conditional independence model with interactions included to represent the associations between location and accident type and environment and accident type. The omitted interaction (LE) indicates that location and environment are independent given accident type.

Readers should note that no further reduction in parameter effects is possible from this model because the two interactions require the presence of all three main effects. In four- or other higher-order tables it may be possible to produce parsimonious models in which only a subset of the main effects or lower-order interactions is needed. For example, in a four-way table composed of variables A, B, C and D, the main effect of D, and all the two-way interactions including D, should be tested as candidates for deletion if ABC is the only three-way effect found to be important. The procedures needed to implement this type of testing are set out in Aitkin (1980) and Wrigley (1985).

9.4.5 Screening procedures for interactions

An effective alternative procedure to STP which may be used to identify the parsimonious log-linear model is the screening strategy suggested by Brown (1976, 1981) and Benedetti and Brown (1978). Once again, a baseline model is used in a stepwise testing of terms. However, unlike STP, these terms are not tested simultaneously in families, but are tested individually against two distinct tests of significance. These are:

1 A test of partial association.
2 A test of marginal association.

Both may be applied to a single term using log-linear models which are quite different in form. To illustrate, consider the tests of the two-way interaction between accident type and weather type (ET). The test of partial association of ET involves comparing the effects on deviance reduction of fitting:

$$\%GM+L+T+E+LT+LE \tag{9.10}$$

rather than:

$$\%GM+L+T+E+LT+LE+ET \tag{9.11}$$

Partial association thus assesses the significance of ET by comparing the values for deviance associated with a model containing all two-way effects, and a model excluding ET. Conversely, the test of marginal association involves comparing the relative performances of:

%GM+E+T+ET   (9.12)

and

%GM+E+T   (9.13)

In this test, only the specific two-way effect and its lower-order relatives are used.

Terms assessed by these two tests may be classified into one of the following three categories:

1 Terms which are significant according to both tests and which must be included in the final log-linear model.
2 Terms which are insignificant according to both tests and which must be excluded from the final log-linear model.
3 Terms whose assessment differs on each test.

Terms which are allocated to category 3 fall into a grey area for which no clear-cut guidelines exist. They may be excluded, but then again, they may also be included. Researchers may have to appeal to theory or rely on intuition for further help.

Table 9.24 contains the results of applying screening to the data in Table 9.18. The probability values listed in the table indicate the probability of rejecting each term from the model. Large values for these probabilities indicate terms which may be eliminated, small values, terms which should not be eliminated. From this table it seems clear that the three-way interaction (LET) and the two-way interaction between location and environment should be eliminated. The two-way interaction between environment and accident type has a small but measurably higher probability value and so is a potential candidate for exclusion. The decision to delete or include it is a matter of personal judgement. The final model suggested by screening is thus very similar to that produced by the simultaneous testing procedure. (Note that the figures for marginal and partial association for main effects and the higher-order interaction in the model are always the same. This is the reason for the partial printing of the figures in Table 9.24.)

9.4.6 Residual analysis

The main thrust of the analysis so far has been to find a log-linear model which may represent the key features of the observed data in as few parameters as possible. The guide to this has been the measure of

Table 9.24 Partial association and marginal association
screening tests

Term	df	Partial association Chi-square	Prob	df	Marginal association Chi-square	Prob
E	2	236.67	0.0			
T	1	46.60	0.0			
L	2	58.73	0.0			
ET	2	7.78	0.0204	2	7.79	0.0204
LE	4	2.76	0.5988	4	2.77	0.5974
LT	2	12.76	0.0017	2	12.76	0.0017
LET	4	4.53	0.3389			

Note: Both tests are identical when applied to main effects or the
highest-order interaction

deviance, a measure of the global or overall fit of the model to the observed data. However, as we saw in Chapter 8, it is not sufficient to rely on global measures. More detailed information on the local fit of a model may be obtained by looking at the residuals from the parsimonious model. As with regression residuals, these may highlight parts of the table where the model fits less well, indicating that further analysis may be required to understand the data better.

There are several different types of residual which may be defined for log-linear models. These differ in their complexity, their assumed distributional properties, and in the sorts of information they may be able to provide. As before, we may define simple residuals (RES) by subtracting the expected cell frequency from the observed cell frequency in each cell. Alternatively, standardised (SR) and adjusted (ADJ) residuals may be calculated. (McCullagh and Nelder 1983 describe a number of other types – Anscombe residuals, deviance residuals and Freeman–Tukey residuals.) In terms of GLIM notation, the three types of residuals described in Chapter 8 may be calculated as follows:

RES N − %FV
SR   (N − %FV)/%SQRT(%FV)
ADJ (N − %FV)/%SQRT(%FV * (1 − %FV * %VL))

where

N   refers to the observed cell frequency
%FV refers to the expected cell frequency, and
%VL refers to the estimated variances of the linear predictors

Further information on their calculation may be found in Defize (1980) and Gilchrist (1981). (Note that the formulae for the standardised and adjusted residuals are slightly different from those outlined in Chapter 8. This is because the specification of Poisson rather than Normal errors requires a modification in the formulae.) Table 9.25 lists the three

Table 9.25 Three types of log-linear residual

N	RES	SR	ADJ
28.00	−0.4768	−0.8935E−01	−0.1660
1.000	−1.563	−0.9763	−1.087
14.00	2.040	0.5898	0.7496
20.00	−1.675	−0.3598	−0.6096
5.000	1.866	1.054	1.275
16.00	−0.1911	−0.4748E−01	−0.7101E−01
97.00	4.947	0.5156	1.208
9.000	0.7152	0.2485	0.3487
33.00	−5.662	−0.9106	−1.459
41.00	1.879	0.3004	0.6017
6.000	0.3439	0.1446	0.2064
27.00	−2.223	−0.4112	−0.7270
75.00	−4.470	−0.5014	−1.111
8.000	0.8477	0.3170	0.4210
37.00	3.623	0.6270	0.9506
22.00	−0.2038	−0.4325E−01	−0.7358E−01
1.000	−2.210	−1.234	−1.498
19.00	2.414	0.5928	0.8901

Notes: N = observed cell frequency
RES = simple residuals
SR = standardised residuals
ADJ = adjusted residuals

types of residual associated with the best-fitting log-linear model found previously. The standardised residuals are all reasonably small with none exceeding plus or minus 2, the value used to indicate potentially extreme residuals.

9.4.7 Influential observations and cells

The use of residuals to guide analysis is an important component in the analysis of cross-classified data, particularly if it helps to identify cells, or groups of cells, whose performance seems to be at odds with the rest of the table. In a similar way, it is also valuable to check a table to see if any cells or groups of cells appears to exert an undue influence on the overall measures of fit. Leverage statistics may be used in exactly the same way as for regression to identify influential cells.

9.5 INCOMPLETE AND RESTRICTED CELLS

All of the contingency tables described so far have been complete and fully-nominal in design. The term complete refers to the fact that none of the observed cell frequencies have been restricted to a fixed value by the researchers (or theory) prior to the analysis. In some circumstances, however, it is possible (and logically necessary) to apply some such restriction, the most usual being that the observed value in the specified cell be set at zero. Whenever there are zero or restricted cells the basic rules of modelling, and, in particular, the calculation of the degrees of freedom, may need to be amended. The procedures outlined in the previous section really apply to complete tables: tables in which every cell in the observed table contains a non-zero value.

If a contingency table contains zero observed cell frequencies, it is important to find out why. Two distinct types of zero cell may be recognised:

1 *Sampling zeros*: cells which exhibit zero frequencies merely because combinations of category levels were not observed in the sample data.
2 *Structural zeros*: cells whose frequencies have been constrained at zero by design (that is, the cell values cannot be other than zero or a predetermined non-zero value).

Constrained cells containing non-zero observed frequencies may also be met in practice. These are treated in exactly the same way as for structural zeros.

9.5.1 Sampling zeros

A table containing sampling zeros is not restricted, but it is worth describing some of its characteristics because it may cause problems in analysis. Table 9.26 presents an extract of some data collected among Ugandan refugees in Sudan (Harrell-Bond 1986). It contains three zero entries for children who are not malnourished. These entries are zero because no incidences of these combinations were evident in the children surveyed. As there is no logical reason why these combinations should be zero, it is possible to argue that additional study in the field might have generated some non-zero values.

Table 9.26 Example of an incomplete contingency table (sampling zero cells)

Guardian's psychological state	Child's health	
	NM	CM
Acutely anxious and severely depressed	0	1
Acutely anxious	0	1
Severely depressed	0	7
Mildly anxious and/or depressed	12	14
Good mental health	17	3

Note: NM: not malnourished
CM: clinical malnourishment
Source: Harrell-Bond (1986)

267

Printout 9.4 Transcription of GLIM analysis of Table 9.26

```
£r *glim
  $UNITS 10
  $DATA N
  $READ
  0 1 0 1 0 7 12 14 17 3
  $FAC CH 2 PH 5
  $CALC CH=%GL(2,1):PH=%GL(5,2)
  $LOOK N CH PH$
```

1	0.0	1.000	1.000
2	1.000	2.000	1.000
3	0.0	1.000	2.000
4	1.000	2.000	2.000
5	0.0	1.000	3.000
6	7.000	2.000	3.000
7	12.00	1.000	4.000
8	14.00	2.000	4.000
9	17.00	1.000	5.000
10	3.000	2.000	5.000

```
  $YVAR N
  $ERR P$
  $LINK LOG$

  $FIT +CH.PH$
          SCALED
  CYCLE DEVIANCE      DF
    10   0.1362E-03    0
  ──── NO CONVERGENCE BY CYCLE 10

  $D ERM$
```

	ESTIMATE	S.E.	PARAMETER
1	−10.69	127.3	%GM
2	−0.3638E−06	180.0	PH(2)
3	−0.3639E−06	180.0	PH(3)
4	13.18	127.3	PH(4)
5	13.53	127.3	PH(5)
6	10.69	127.3	CH(2)
7	0.3638E−06	180.0	PH(2).CH(2)
8	1.946	180.0	PH(3).CH(2)
9	−10.54	127.3	PH(4).CH(2)
10	−12.43	127.3	PH(5).CH(2)

SCALE PARAMETER TAKEN AS   1.000

UNIT	OBSERVED	FITTED	RESIDUAL
1	0	0.2270E−04	−0.4764E−02
2	1	1.000	0.3351E−13
3	0	0.2270E−04	−0.4764E−02
4	1	1.000	0.2371E−10
5	0	0.2270E−04	−0.4764E−02
6	7	7.000	0.2523E−05
7	12	12.00	0.3304E−05
8	14	14.00	0.0
9	17	17.00	0.3932E−05
10	3	3.000	0.0

```
Y-VARIATE N
ERROR POISSON LINK LOG
LINEAR PREDICTOR
%GM PH CH PH.CH
$STOP$
```

Note: Commands prefixed by £ are operating system commands
Commands prefixed by $ are GLIM commands

The analysis of a table with zero observed frequencies mirrors that of the complete table. In the case of Table 9.26 the analysis is almost identical because the empty cells are a minority of the cells in the table, and they are not arranged in a distinctive pattern. As a result, it is possible to fit a full range of unsaturated log-linear models to the table, each of which produces non-zero estimates for every cell, including the three which have zero observed frequencies. For example, under independence, the expected values for cells 11 and 21 are both 5.273, and for cell 31, 3.691.

However, difficulties arise if researchers attempt to fit the saturated model. For this model, the presence of the zero cells prohibits the calculation of the expected cell frequencies because information on each observed frequency is used in their calculation. As the natural logarithm of zero is infinity estimates cannot be produced. Printout 9.4 contains a transcription of the GLIM analysis of Table 9.26 which illustrates the basic problem. Notice that after ten iterations the program still had not converged. Though expected cell frequencies similar to those in the observed table have been produced, the estimation procedure has not been completed.

One way of removing the problem of the sampling zero cells is simply to replace the expected cell frequencies by the observed cell frequencies and not to rely on an estimation algorithm. This is satisfactory in itself but there are alternative procedures which may be used. One popular approach is to replace the zero cells by a small positive value, for example, 0.5. Printout 9.5 illustrates what happens when 0.5 is added to the zero cells in Table 9.26. First, the algorithm has no difficulty fitting the saturated model. Second, the values for deviance are deflated in the modified analysis compared with the original analysis. This deflation is generally not sufficient to lead to fundamentally different interpretations of the patterns in the data.

One of the main drawbacks to the addition of 0.5 is that it is a wholly arbitrary practice. Though Goodman (1970) argues that 0.5 should be added to every cell regardless of whether it is zero or not, this, in itself, is arbitrary. Moreover, it may be a source of confusion if the cell frequencies being modified are naturally small as, for example, when representing the number of major floods over a period in particular areas. If a value is to be added, it only seems reasonable that this should be a value suggested

Printout 9.5 Re-analysis of Table 9.26

```
$r *glim
  $UNITS 10
  $DATA N
  $READ
  0 1 0 1 0 7 12 14 17 3
  $FAC CH 2 PH 5
  $CALC CH=%GL(2,1):PH=%GL(5,2)
  $CALC N=N+0.5
  $YVAR N
  $LINK LOG
  $ERR P
  $FITS$
```

	SCALED	
CYCLE	DEVIANCE	DF
4	65.21	9

`$FIT CH+PH$`

	SCALED	
CYCLE	DEVIANCE	DF
4	18.58	4

`$FIT +CH.PH$`

	SCALED	
CYCLE	DEVIANCE	DF
4	0.2918E−10	0

`$D ER$`

	ESTIMATE	S.E.	PARAMETER
1	−0.6931	1.414	%GM
2	1.099	1.633	CH(2)
3	−0.4340E−07	2.000	PH(2)
4	−0.1878E−07	2.000	PH(3)
5	3.219	1.442	PH(4)
6	3.555	1.434	PH(5)
7	0.4101E−06	2.309	CH(2).PH(2)
8	1.609	2.191	CH(2).PH(3)
9	−0.9502	1.678	CH(2).PH(4)
10	−2.708	1.735	CH(2).PH(5)

SCALE PARAMETER TAKEN AS   1.000

UNIT	OBSERVED	FITTED	RESIDUAL
1	1	0.5000	−0.4215E−07
2	2	1.500	0.5110E−06
3	1	0.5000	0.0
4	2	1.500	0.0
5	1	0.5000	0.0
6	8	7.500	0.2612E−05
7	13	12.50	0.3372E−05
8	15	14.50	0.3631E−05
9	18	17.50	0.3990E−05
10	4	3.500	0.0

`$STOP$`

Note: Commands prefixed by £ are operating system commands
Commands prefixed by $ are GLIM commands

by prior experience or some given theory. Such an approach is a form of pseudo- or empirical-Bayesian analysis, and has much to commend it in that information obtained in other situations may be incorporated directly into the analysis. However, unlike the data reduction procedures proposed by Ehrenberg (1982) and outlined in Part I, it is not immediately clear how these theoretical values may be obtained. As was mentioned in Chapter 5, the main problem with Bayesian analysis is the calculation of the prior probabilities (or observed values in this context). Bishop *et al.* (1975: Chapter 10) and Wrigley (1985) discuss the matter in more detail.

9.5.2 Empty marginals

Table 9.27 is a second example of a table containing sampling zero cells. This differs from Table 9.26 in that the empty cells are positioned in such a way that a marginal total of the table is also zero. In order to fit a full-set of log-linear models, all marginals of a contingency table must possess non-zero values so that their parameters may be estimated.

Table 9.27 Sampling incompleteness: empty marginal

		TYPE					
		S ENV			M ENV		
		D	R	N	D	R	N
	Far	28	1	14	20	0	16
LOCATION	Medium	97	9	33	41	0	27
	Near	75	8	37	22	0	19

Note: TYPE: S = single vehicle/pedestrian accidents
M = multiple vechicle accidents

ENV(IRONMENT): D = Daytime accidents
R = Daytime accidents on rainy days
N = Night-time accidents
LOCATION: Distance to nearest hospital

The main complication encountered in analysing this type of incomplete table is that the expected cell frequencies associated with the incomplete marginal total must always equal zero. This restriction arises because the table does not contain the necessary information to provide non-zero expected cell frequencies for those cells. However, non-zero expected frequencies may be generated for them by log-linear models which do not include the incomplete marginal in their calculations. If an attempt is made to fit the incomplete marginal, a loss of degrees of freedom occurs and the estimating algorithm cannot converge. Neither the $CYCLE nor $RECYCLE commands are of help here because the problem lies with a lack of information rather than convergence limits.

This lack of information is reflected in the calculation of the degrees of freedom for the table. Because the empty cells do not contain independent information it is important to delete them from the calculation of the degrees of freedom for appropriate models. Bishop *et al.* (1975) and Fienberg (1977) show that the correct degrees of freedom for a model applied to a table containing zero marginals may be estimated from the following formula:

$$DF = (TC - EC) - (TP - EP) \tag{9.14}$$

where

    TC refers to the total number of cells in the table

    EC refers to the total number of empty cells

    TP refers to the total number of parameters to be included in the model, and

    EP refers to the total number of parameters which cannot be fitted because of the empty marginal

This may be automated within GLIM using a macro (Payne 1979) – a subset of commands which may be activated as required from any point of the main program. The key feature of Payne's approach is to use a weight vector to eliminate zero cells which form part of a zero marginal total. When activated, the macro checks the data for missing cells and estimates the correct degrees of freedom as required. If the data are satisfactory in the first place, the macro does not proceed beyond its initial checking. Printout 9.6 illustrates the procedure. Notice that the macro (CDF) is called on several occasions but is only activated when the two-way interaction between environment and accident type is fitted to the mutual independence model. The degrees of freedom are reduced from ten to eight as a result though the deviance value remains unchanged.

The macro illustrates a number of additional features of the GLIM language. First, the use of the dot operator in the $FIT command ($FIT .$). This causes GLIM to refit the previous model and removes the need to type the terms a second time. Second, the use of the %GT and %EQ operators to check for the equality or relative magnitude of two variables. Third, %CU is used to produce a cumulative total of the values in WTT. This operator is particularly useful in generating totals or calculating marginal values for data tables. Finally, the $EXIT command provides a mechanism to pass control back to the main program if adjusted degrees of freedom are identical to those produced in the main program. Printing from the macro only occurs if adjusted degrees of freedom have been calculated.

9.5.3 Structural zeros

A contingency table containing structural zero cells is illustrated in Table 9.28. This shows the number of corporate interlocks (closed networks of

272

Printout 9.6 Use of a macro to calculate the correct degrees of freedom

```
$R *GLIM
  $MACRO CDF
  $CALC WTT=%GT(%FV,0.001)*WT
  $CALC %T=%EQ(%CU(WTT),%NU)
  $EXIT %T
  $PRINT '*** CORRECT DF FOLLOWS ****'
  $WEIGHT WTT $FIT.$
  $WEIGHT WT
$ENDMAC
$UNITS 18
$DATA N
$READ
28 1 14 20 0 16
97 9 33 41 0 27
75 8 37 22 0 19
$FACL 3 E 3 T 2
$CALCL=%GL(3,6):E=%GL(3,1):T=%GL(2,3)
$CALC WT=1
$WEIGHT WT
$YVAR N
$LINK LOG
$ERR P
$FIT$
```

	SCALED	
CYCLE	DEVIANCE	DF
5	433.7	17

```
$FIT L+E+T$
```

	SCALED	
CYCLE	DEVIANCE	DF
4	35.45	12

```
$FIT +E.T$
```

	SCALED	
CYCLE	DEVIANCE	DF
9	13.65	10

```
$USE CDF$
```

*** CORRECT DF FOLLOWS ****
————— CURRENT DISPLAY INHIBITED

	SCALED	
CYCLE	DEVIANCE	DF
3	13.65	8

————— CURRENT DISPLAY INHIBITED

```
$STOP$
```

Note: Commands prefixed by £ are operating system commands
Commands prefixed by $ are GLIM commands

Table 9.28 A structurally incomplete contingency table

| | Number of interlocks | | | | |
Location	16+	6–15	2–5	1	0
Johannesburg	11	45	37	8	14
Cape Town	0	0	16	37	62
Durban	0	0	17	29	69
Stellenbosh	0	0	10	9	96
Pretoria	0	0	13	17	75
Others	0	0	19	18	78

Source: Based on Cox and Rogerson (1985)

information and power) which exist in the South African business community. Because of the nature of corporate business, the number of interlocking directorships which can occur in a particular locality depends on the business concerned and the relative importance of the locality. From Table 9.28, it is clear that major information networks may only occur in Johannesburg. The zero cells are therefore restricted in that additional sampling will not reveal their presence outside Johannesburg.

There are a number of ways of analysing a structurally-incomplete table. The simplest approach is to rewrite the table so that the non-restricted cells are gathered together into complete squares or rectangles. This may be done for Table 9.28 producing three distinct structures:

1 The complete 6 row by 3 column table of small interlocks.
2 The complete 2 column row vector associated with major interlocks in Johannesburg.
3 The 5 row by 2 column block of zero cells.

The ability to rewrite a table in this way arises because Table 9.28 is said to be separable, that is, capable of being decomposed into distinct complete subsets. The two subsets containing the non-restricted data may thus be analysed separately, and their degrees of freedom combined to produce the total for the full table. Under independence there are zero degrees of freedom associated with the row vector and ten degrees of freedom with the complete table for small interlocks. The total for the table as a whole is thus ten degrees of freedom. By extension, models other than independence may also be applied to two subsets.

Unfortunately, contingency tables are rarely separable and so it is not always possible to re-express the data in these ways. Given inseparability, attention must focus on analysing the table as it is actually written. However, this requires the modification of our concepts of independence and interaction as these have been devised explicitly for complete tables. One important modification suggested by Goodman (1968) uses the concepts of quasi-independence and quasi-log-linear models. These are variants of the procedures outlined earlier which apply to the cells in the non-restricted rows and columns of a structurally-incomplete table.

The calculation of the degrees of freedom for quasi-log-linear models differs from the procedures outlined earlier for either the complete table or the incomplete table with zero marginals. Instead of using the standard formulae, the following expression should be used:

$$DF = TC - RC - TP \qquad (9.15)$$

where

TC and TP are as before, and
RC refers to the total number of restricted cells in the table

For Table 9.28, TC=30, RC=10 and TP under quasi-independence equals 10. This means that the quasi-log-linear model for quasi-independence has ten degrees of freedom, identical to the number calculated by analysing the components of Table 9.28 separately. The concepts of quasi-independence and the quasi-log-linear model generalise in exactly the same way as their counterparts for the complete table, providing a powerful family of techniques which are suitable for the restricted design.

9.5.4 Square tables

Contingency tables which are square may frequently be created by, for example, measuring some characteristic of a sample of individuals at two points in time. The row classifications may thus correspond to the patterns observed in the former period and the column classifications to the latter period. For designs of this nature, a key form of analysis is the assessment of the stability of the classifications through time. In particular, researchers may be keen to know under what circumstances classifications may change through time rather than remain stable.

The analysis of square tables is particularly important in research in public health, medicine, educational research and psychology, where subjects are frequently subjected to repeated measurements to assess their state of health, knowledge or reaction time. Similarly, in sociology, studies are frequently made to assess occupational and social mobility. These involve the measurement and comparison of the occupations of parents and children and may be used in studies of social cohesion.

Square tables are rather more complex to analyse than the contingency tables presented previously because the diagonal elements frequently behave differently from the remainder of the table. These elements correspond to the number of observations in the sample which retain their original classification through time. A variety of different types of procedure based on odds-ratios may be used to compare these elements with the off-diagonal elements (Duncan 1979; Goodman 1979a). Many of these may be specified within GLIM using log-linear models. A typical mode of analysis is to treat the square table as a structurally incomplete table in which the elements on the principal diagonal are eliminated from the model. This may be done using a weight variable in which the row and column classifications are compared, and if identical, eliminated. Thus if A refers to the row classification and B to the column, then:

$CALC WT=%NE(A,B)
$WEIGHT WT

would eliminate all cells from the model fit if their row and column classification levels were equal, that is cells 11, 22, 33, etc. Jones and Pittelkow (1983) present a number of GLIM macros which may be of help in analysing square tables. Similarly, Weber (1981) contains a series of macros for assessing two hypotheses which are particularly relevant to the square table, marginal homogeneity and symmetry. (Further details of these and other features of the square design may be found in Bishop *et al.* 1975.)

9.6 ASYMMETRIC TABLES

In the previous sections the object has been to investigate interrelationships and interdependencies within a contingency table. We have seen that models suited to this form of analysis may be fitted in GLIM by (a) specifying the expected cell frequency under some hypothesis as the response or dependent variable, (b) declaring that the cell frequencies follow independent Poisson distributions, and (c) linking the predictable mean of the response variable to the linear predictor using a logarithmic link. The number of terms and parameters which may be included in this general symmetric analysis depends on the shape and structure of the table in question, in particular, on the number of cross-classifying dimensions present, and on restrictions applying to particular cells or marginals.

An alternative type of contingency table analysis arises when the response variable is defined to be the expected cell frequencies in a given dimension of the table rather than the table as a whole. Attention then focuses on how these frequencies vary given the classifications of the remaining dimensions of the table. These dimensions play a similar role to the independent or explanatory variables in a regression model. Because interest is centred on dependency relationships in one dimension, this type of tabular analysis is frequently termed asymmetric.

Two distinct approaches may be used to analyse asymmetric dependency relationships in contingency tables. The first is a generalisation of the standard log-linear approach using Poisson errors and logarithmic link. The second is appropriate if the table of observed frequencies is rewritten as a table of observed proportions. This may then be analysed using binomial errors and a logit link. The model which GLIM fits to this configuration of commands is termed a logit regression model.

9.6.1 Asymmetric log-linear models

The use of log-linear models to reproduce dependency relationships in contingency tables relies on the fact that the calculation of the expected cell frequencies under different hypotheses makes use of specific combinations of observed marginal totals. For example, the expected frequencies of a two-way table under independence are calculated directly from the observed row and column marginals which are preserved in the table of expected cell frequencies. Under different hypotheses, only some of these observed marginals will be preserved. This preservation arises because of the definition of the terms being fitted to the log-linear model and is not the result of an explicit decision on the part of the researchers to fix the marginals in advance. However, it is perfectly possible for these marginal totals to be fixed in advance. If this is the case, then models fitted to the table which attempt to describe patterns in the data must reproduce both the grand total and the fixed marginals. If the log-linear model terms corresponding to these marginal totals are included in every model, including those which seem to be insignificant according to screening or STP, then the fixed marginals will always be reproduced. This feature underlies the use of log-linear models in analysing the asymmetric table.

This approach may be illustrated using the multi-way data in Table 9.29 which extends the information in Table 9.18 into four dimensions by identifying the number of times an ambulance was called for each combination of location, environment and accident type. If we assume that the binary variable (RESP) is the response variable then the following log-linear model is the simplest which may be fitted which reproduces the observed marginal totals:

%GM+LOC*TY*ENV

The * notation is a shorthand used within GLIM to include all possible combinations of LOC, TY and ENV. Thus this model may be written out fully as:

%GM+LOC+TY+ENV+LOC.TY+LOC.ENV+TY.ENV
+LOC.TY.ENV   (9.16)

This model may be extended in stages to include the main effect of RESP (base model + RESP), all two-way interactions between the explanatory variables and RESP, all three-way interactions with RESP, and finally, the single four-way interaction for the table. The effect of these additions on deviance is summarised in Table 9.30 and Printout 9.7.

Table 9.29 An asymmetric contingency table

				TYPE					
				S			M		
				ENV			ENV		
				D	R	N	D	R	N
			Far	9	0	7	11	3	10
	Yes	LOC	Medium	30	5	12	13	3	15
			Near	13	3	12	2	1	8
RESPONSE									
			Far	19	1	7	9	2	6
	No	LOC	Medium	67	4	21	28	3	12
			Near	62	5	25	20	0	11

Note: For a list of the abbreviations see Table 9.18

Table 9.30 Analysis of deviance table for log-linear analysis of
Table 9.29

Model	Deviance	DF	Change Deviance	DF
%GM	458.2	35		
RESP+LOC*TY*ENV	41.8	17	416.4	18
+ 2-way	9.8	12	32.0	5
+ 3-way	4.4	4	5.4	8
+ 4-way	0.0	0	4.4	4

9.6.2 Logit models for contingency tables

The second approach which may be used to analyse dependency in a contingency table is to fit logit models rather than log-linear models. Logit is a term used to refer to a particular type of analytic transformation which may be applied to a dependent variable consisting of proportions (see Cox 1970 for further details). Instead of using a logarithmic link to relate the frequency data to the linear predictor as in log-linear models, a logit link is used to relate the proportions data instead. These two distinct approaches may be written out algebraically as:

log-linear models: $\eta = \ln(\mu)$
logit models: $\eta = \ln(\mu/(N-\mu))$

Table 9.31 presents the accident data in the form of proportions.

Printout 9.8 summarises the logit analysis of these data. The following differences should be noted when comparing this with the log-linear approach:

1 The $UNITS command is set to 18 rather than 36 to reflect the fact that a pair of figures is to be inserted for each observation. The first figure is the number of times an ambulance was called, the second, the total

Printout 9.7 Asymmetric log-linear analysis of Table 9.29

```
£r *glim
  $UNITS 36
  $DATA N
  $READ
  9 19 11 9 0 1 3 2 7 7 10 6 30 67 13 28 5 4 3 3
  12 21 15 12 13 62 2 20 3 5 1 0 12 25 8 11
  $FAC RESP 2 LOC 3 ENV 3 TY 2
  $CALC RESP=%GL(2,1):LOC=%GL(3,12):ENV=%GL(3,4):TY=%GL(2,2)
  $LOOK N RESP LOC ENV TY$ (list truncated)
```

1	9.000	1.000	1.000	1.000	1.000
2	19.00	2.000	1.000	1.000	1.000
3	11.00	1.000	1.000	1.000	2.000
4	9.000	2.000	1.000	1.000	2.000
5	0.0	1.000	1.000	2.000	1.000
6	1.000	2.000	1.000	2.000	1.000
7	3.000	1.000	1.000	2.000	2.000
8	2.000	2.000	1.000	2.000	2.000
9	7.000	1.000	1.000	3.000	1.000
10	7.000	2.000	1.000	3.000	1.000
11	10.00	1.000	1.000	3.000	2.000
12	6.000	2.000	1.000	3.000	2.000

```
  $YVAR N $LINK LOG$ERR P$
  $FIT$
            SCALED
  CYCLE  DEVIANCE  DF
     5     458.2    35

  $FIT RESP+LOC*TY*ENV$
            SCALED
  CYCLE  DEVIANCE  DF
     4     41.81    17

  $FIT +RESP*LOC +RESP*ENV +RESP*TY$
            SCALED
  CYCLE  DEVIANCE  DF
     4     9.757    12

  $FIT +RESP*LOC*ENV +RESP*LOC*TY +RESP*ENV*TY$
            SCALED
  CYCLE  DEVIANCE  DF
     4     4.400     4

  $FIT +RESP*LOC*TY*ENV$
            SCALED
  CYCLE  DEVIANCE  DF
     2    0.9080E-04  0

  $STOP$
```

Note: Commands prefixed by £ are operating system commands
Commands prefixed by $ are GLIM commands

279

Table 9.31 Observed proportions (logit analysis)

				____	*S*	____	____	*TYPE* M	____
					ENV			*ENV*	
				D	*R*	*N*	*D*	*R*	*N*
			Far	9	0	7	11	3	10
	Yes	*LOC*	*Medium*	30	5	12	13	3	15
			Near	13	3	12	2	1	8
RESPONSE									
			Far	28	1	14	20	5	16
	Total	*LOC*	*Medium*	97	9	33	41	6	27
			Near	75	8	37	22	1	19

Notes: For a list of the abbreviations see Table 9.18

number of accidents in each cross-classification of the explanatory variables.

2 Two variables are defined in the $READ command. The first corresponds to the numerator of the observed proportion, the second, to its denominator.
3 The $YVAR command specifies that RESP is to be treated as the response rather than the *N* frequencies of Table 9.29.
4 The Poisson error process of the symmetric analysis is replaced by a binomial process (B). Notice that TOT is included in the specification of the error process. This is needed to define the denominator term explicitly.
5 A logit link ($LINK G) is used to relate the linear predictor to the predictable mean rather than the logarithmic link of the symmetric analysis.

Table 9.32 displays the effect of this logit model on deviance. The simplest model is the grand mean effects model (%GM). This yields a deviance value of 41.8 for seventeen degrees of freedom. This is equivalent to an hypothesis of constant logit with each expected cell frequency being identical in value. Notice that this corresponds to the log-linear model in which RESP has been added to the base model. The addition of the three main effects reduces deviance to 9.8 for twelve degrees of freedom. This is equivalent to the log-linear model in which all two-way interactions including RESP have been added to the model. The addition of the logit two-way effects reduces deviance to 4.4 for four degrees of freedom. This is equivalent to the log-linear model in which all three-way effects including RESP have been added to the base model. Finally, the addition of the logit three-way effects produces the saturated logit model. This is equivalent to the log-linear model which includes the four-way interaction between all RESP, ENV, TY and LOC.

Printout 9.8 Asymmetric logit analysis of Table 9.31

```
£r *glim
  $UNIT 18
  $DATA RESP TOT
  $READ
  9 28 11 20 0 1 3 5 7 14 10 16 30 97 13 41 5 9
  3 6 12 33 15 27 13 75 2 22 3 8 1 1 12 37 8 19
  $FAC LOC 3 ENV 3 TY 2
  $CALC LOC=%GL(3,6):ENV=%GL(3,2):TY=%GL(2,1)
  $LOOK RESP TOT LOC ENV TY$
```

	RESP	TOT	LOC	ENV	TY
1	9.000	28.00	1.000	1.000	1.000
2	11.00	20.00	1.000	1.000	2.000
3	0.0	1.000	1.000	2.000	1.000
4	3.000	5.000	1.000	2.000	2.000
5	7.000	14.00	1.000	3.000	1.000
6	10.00	16.00	1.000	3.000	2.000
7	30.00	97.00	2.000	1.000	1.000
8	13.00	41.00	2.000	1.000	2.000
9	5.000	9.000	2.000	2.000	1.000
10	3.000	6.000	2.000	2.000	2.000
11	12.00	33.00	2.000	3.000	1.000
12	15.00	27.00	2.000	3.000	2.000
13	13.00	75.00	3.000	1.000	1.000
14	2.000	22.00	3.000	1.000	2.000
15	3.000	8.000	3.000	2.000	1.000
16	1.000	1.000	3.000	2.000	2.000
17	12.00	37.00	3.000	3.000	1.000
18	8.000	19.00	3.000	3.000	2.000

```
  $YVAR RESP$ERR B TOT$LINK G$
  $FIT$
              SCALED
    CYCLE  DEVIANCE  DF
        3    41.81    17

  $FIT +LOC$
              SCALED
    CYCLE  DEVIANCE  DF
        3    27.02    15

  $FIT +ENV$
              SCALED
    CYCLE  DEVIANCE  DF
        3    11.97    13

  $FIT +TY$
              SCALED
    CYCLE  DEVIANCE  DF
        3    9.757    12

  $D ER$
    ESTIMATE     S.E.        PARAMETER
    1 −0.5756   0.2590       %GM
    2 −0.3711   0.2680       LOC(2)
    3 −0.9962   0.2948       LOC(3)
```

281

Printout 9.8 Continued

4	0.9308	0.3985	ENV(2)
5	0.7142	0.2202	ENV(3)
6	0.3204	0.2144	TY(2)

SCALE PARAMETER TAKEN AS 1.000

UNIT	OBSERVED	OUT OF	FITTED	RESIDUAL
1	9	28	10.08	−0.4247
2	11	20	8.731	1.023
3	0	1	0.5879	−1.194
4	3	5	3.314	−0.2968
5	7	14	7.484	−0.2595
6	10	16	9.804	0.1004
7	30	97	27.12	0.6523
8	13	41	14.28	−0.4203
9	5	9	4.464	0.3571
10	3	6	3.453	−0.3744
11	12	33	14.59	−0.9080
12	15	27	14.09	0.3495
13	13	75	12.90	0.3144E−01
14	2	22	4.894	−1.484
15	3	8	2.760	0.1783
16	1	1	0.4205	1.174
17	12	37	11.02	0.3523
18	8	19	7.008	0.4718

$FIT +ENV.TY+ENV.LOC+TY.LOC$
      SCALED

CYCLE	DEVIANCE	DF
4	4.400	4

$FIT +LOC.ENV.TY$
      SCALED

CYCLE	DEVIANCE	DF
10	0.9132E−04	4

$STOP$

Note: Commands prefixed by £ are operating system commands
Commands prefixed by $ are GLIM commands

9.6.3 Multi-way response variables

Asymmetry problems involving a multi-level response variable are the natural extension of the binary problems outlined in the previous sections. The log-linear analysis of the multi-way table has already been presented and so should not prove to be problematic. As with the binary case, it is important to remember that the base model must include the main effects of the explanatory variables and any interactions between them. Extensions to this then involve the addition of the main effect of the response, the two-way interactions between the response and the explanatory variables, and so on until saturation. Screening and STP may be used with

Table 9.32 Analysis of deviance table for logit
analysis of Table 9.30

Model	Deviance	DF	Change Deviance	DF
%GM	41.8	17		
+ main effects	9.8	12	32.0	5
+ 2-way	4.4	4	5.4	8
+ 3-way	0.0	0	4.4	4

these models as with the symmetric analyses, except that the terms required to fix the observed marginals associated with the explanatory dimensions must always be included.

The logit analysis of the multi-way table requires that the cell proportions follow a multinomial rather than a binomial distribution. This is currently not available within GLIM and so it is not entirely feasible to specify a multivariate logit model. However, Goldstein (1979) notes that it is possible to generalise the binomial distribution so that comparisons between levels in a multi-way response variable may be made. This amounts to the specification of a multivariate logit model.

9.7 CATEGORICAL REGRESSION

The notion of asymmetric analysis involving categorical data extends beyond the confines of data tables. Categorical measurements may frequently be used as the response variables in regression-type models. For a variety of technical reasons, the classic regression model as introduced in Chapter 8 cannot always be used when the response variable is categorical. Wrigley (1976, 1985) notes that categorical response variables in traditional regression models not only violate the assumption of constant error variance but may also generate predicted values for the response variable which are uninterpretable. However, Bartholomew (1981) shows that there are a number of specific types of situation in which neither of these difficulties arise and so the ordinary least squares procedures of classic regression may be applied.

The solutions offered to remove the perceived difficulties of categorical responses involve transforming them so that new linear additive models are specified instead of classic regression. The family of folded root transformations provides a number of possibilities (Cox 1970), two of which lead to tractable models. These are the cumulative logistic transformation which leads to the logit model, and the cumulative Normal distribution which leads to the probit model. The logit regression model for non-tabular data is a generalisation of the model described previously. Both of these models may be specified with ease within GLIM for analyses involving binary response variables. To illustrate their use, the precipitation data presented earlier are re-analysed. However, for the purpose of this example the precipitation values in variable PPT have been recoded to two non-overlapping class levels:

1 for values exceeding 20 inches
0 for values less than 20 inches

As with the logit analysis of the asymmetric contingency table, a denominator term is required in the specification of the error command. This was easily solved for the table problem merely by using the cross-classification totals. However, no such totals exist for a non-tabular data set.

This problem may be solved quite easily because the estimated values from a categorical regression model are usually interpreted as predicted probabilities. The denominator may thus be set to 1, indicating the upper limit of probabilities, for each of the thirty binary measurements of PPT. In order to make use of this, a vector, DEN, needs to be created within GLIM to hold the denominator terms. This may be done using the $CALC command:

$CALC DEN=%GL(1,30)

Both the logit and probit models assume binomial errors. Consequently, the error component in both specifications is $ERR B DEN. However, they differ in the links they use. For logit, $LINK G is specified. However, for probit, $LINK P is used. This relates the predictable mean of the response to the linear predictor as follows:

$$\eta = \phi^{-1}(\mu/n) \tag{9.17}$$

In spite of these differences, the two models behave almost identically, both in their estimation of deviance and in the probability values they estimate for each of the thirty weather stations. As with the classic regression model and the log-linear models, no categorical regression analysis is complete without an inspection of the residuals and leverage values. The procedures of use here are identical to those used previously. (For a detailed account of the use of diagnostic procedures with categorical regression models, see Dunn and Wrigley 1984.)

The fact that the two models behave very similarly suggests that they may be substitutes for each other. This is frequently possible and the choice usually depends on which of the two is easier to compute. In general, logit is the simpler and more tractable. However, for many other types of problem their performances differ greatly and the choice may depend on theoretical requirements rather than expediency.

There are many examples of the use of these models in the published applied statistics literature. Logit is particularly widely used in psychology as a means of assessing the reactions to stimulus–response experiments.

Conversely, probit is much more widely used in the biostatistics literature. The main reason for this is that researchers claim the existence of a direct relationship between the cumulative Normal distribution which underlies probit and the theoretical tolerance distributions which underlie bioassay and toxicological research. (For a comparative review of the use of these models in biology, see Hewlett and Plackett 1978.) Geographical examples include, among others, Lewis (1975), Odland and Balzar (1979), Wrigley (1980, 1981), Guy and Broom (1983), O'Farrell and Crouchley (1984) and Stapleton-Concord (1984).

In Chapter 10 some mention will be made of an alternative use of these categorical regression models which may be of interest to some geographers. This is in connection with the use of so-called discrete choice models: theoretical models built upon the use of the logit and probit transformations which are widely used in transport research and in some branches of economics to study individual consumer choices and preferences. Both models offer alternative frameworks for describing some of the characteristics of individual decision-making, allowing hypotheses to be tested and planning alternatives assessed.

9.8 ORDINAL DATA

In the various types of categorical data problem described, attention has focused on nominal measurements. In these, the order of the levels in each classification is arbitrary. Thus it makes no difference to the interpretation of the information in a contingency table if the levels in the rows or columns are repositioned, for example, by renumbering column 2 as column 1. The cross-product ratio is particularly important in the interpretation of interactions between nominal categorical variables simply because it too is unaffected by such repositioning. However, there are many types of geographical data where such a modification might seriously change the fundamental patterns of association. For example, in a regional mobility table indexed by area and time, the time dimension cannot be repositioned at will without making a nonsense of the information contained within it. Similarly, any form of data analysis which involves the use of ordinal data is likely to be inadequate if the ordinal nature of the information is ignored.

For many years the analysis of ordinal data either in contingency tables or categorical regression models has been largely untouched by researchers. Most of the methodological developments in the fields of categorical data analysis have been aimed towards nominal measurements. One of the reasons for this is that ordinal measurements lie in a methodological grey area. Some researchers are apt to consider them wholly arbitrary classifications similar to nominal measurements, whilst others argue that they are actually poorly-measured continuous variables (see the

285

discussion in Chapter 3). As a result, their analysis has varied considerably depending on the tendencies of the researchers, with some suggesting the use of log-linear models, and others advocating regression or analysis-of-variance approaches. However, since the 1970s, and particularly since the late 1970s, considerable improvements have been made to develop and popularise a class of techniques which are tailor-made for ordinal data, and in particular, ordinal contingency table data.

The literature on the analysis of ordinal data is now quite extensive with many techniques being suggested in article or bookform (Haberman 1974b; Simon 1974; Agresti 1980, 1984; Goodman 1979b, 1981). The motivation for much of this work seems to be that:

1 Ordinal techniques are more powerful than conventional log-linear-based procedures, and are better suited in testing null hypotheses about table structure.
2 Ordinal techniques are similar to those used in the analysis of continuous data, thus helping to link the grey area more closely with an established area of methodology.
3 Ordinal techniques provide for a wider variety of alternative types of analysis than log-linear procedures and so may help researchers generate more parsimonious models.

Breen (1985) lists the following as examples of the range of model types currently available:

1 A uniform association model (after Goodman 1979b).
2 A 'row' effects model.
3 A column 'effects' model.
4 An additive 'row+column' effects model.

He suggests that the uniform association model may be considered the most basic of the four.

The key difference between the analysis of nominal tables and their ordinal counterparts is that the concept of interaction needs to be amended to incorporate the given order of the cells in the ordinal dimensions. One approach is to suggest that the cells lie along an underlying continuum and should therefore be given scores of some sort to reflect their relative positions. The patterns of association which are identified in the data reflect the assumptions made about this scoring, for example, whether it refers to global patterns of association between the variables at all classification levels, or is in some sense local, referring to associations between contiguous neighbours.

At the moment the analysis of the ordinal table cannot be performed within GLIM merely by specifying default options as is possible with the nominal design. As a result, the models suggested by Breen, and a number of others – models based on continuation odds (Fienberg and Mason 1979), proportional odds and hazards, and a partial likelihood survival model (Cox 1975; Whitehead 1980) – may only be specified using macros. A number of authors have published macros which may be used to specify these models. (For further details, see Breen 1985; Hutchinson 1985.)

9.9 CONCLUSIONS

The aim of this chapter has been to illustrate some of the major categorical models which can be re-expressed as generalised linear models. The three models which today form the core approaches – the hierarchical log-linear model, the logit model, and the probit model – are all special cases of the generalised linear model. Five distinct types of analytical problem associated with their use are catered for as default options in GLIM. These are:

1 The hierarchical log-linear model for symmetric tables.
2 The hierarchical log-linear model for asymmetric tables.
3 The logit model for asymmetric tables.
4 The logit model for non-tabular categorical regression.
5 The probit model for non-tabular categorical regression.

Each is distinguished from the others by differences in the structure of their linear predictors, or by differences in the $YVARIABLE, $ERROR and $LINK commands. These are summarised in Table 9.33.

From this table it is quite clear that different forms of categorical model can be fitted, where desirable, to the same data set. This is made exceedingly easy within GLIM, and is valuable because it offers the potential for making analysis more comprehensive. The didactic value

Table 9.33 Summary of GLIM specifications for categorical generalised linear models

Model	$YVAR	$ERROR	$LINK	Full factor structure
Symmetric log-linear model	Expected cell frequencies	Poisson	Log	Yes
Asymmetric log-linear model	Specified dimension	Poisson	Log	Yes
Tabular logit	Specified dimension	Binomial	Logit	Yes
Logit regression	Specified variable	Binomial	Logit	No
Probit regression	Specified variable	Binomial	Probit	No

offered by GLIM is also valuable because it permits students to fit a variety of model types to data without their having to wade through different computer packages and analytical treatments developed from more than one research tradition.

The five models presented here are 'basic' in the sense that they are applicable to many analytical research problems and are not specific to any one discipline. However, it is possible to extend these models to accommodate slight variations of the basic analytical problems (for example, to cope with multiple logit models rather than binary models), and also to develop entirely new research specialities using them. Some of these developments are considered in the next chapter. These include the development of choice theory models based on logit and probit and the reconciliation of the mathematical approaches to spatial interaction based on a particular type of log-linear model: the Poisson regression model.

10

SOME EXTENSIONS AND SPECIALISED TOPICS

10.1 INTRODUCTION

The two previous chapters have outlined some of the more commonly-used generalised linear models in geography. In this chapter a number of extensions of these common models, as well as some specialised topics based on them, are presented. Three areas are considered:

1 Poisson regression models;
2 models suited to the analysis of designed experiments;
3 discrete choice models.

10.2 POISSON REGRESSION

10.2.1 A simple example

Table 10.1 presents some data taken from Smith (1975), which shows the observed number of customers at a new shopping centre in Sydney related to the distance they have travelled to get there. Smith used this data set to illustrate the characteristics of the classical linear regression model using ordinary least squares estimation. Regression was considered to be suitable because both variables may be treated as continuous, and a dependency relationship linking one with the other is plausible. Using standard least squares procedures (as outlined in Chapters 6 and 8), the following relationship can be deduced for the regression of centre patronage (Y) on the natural logarithm of distance travelled (X):

$$Y = 201.7 - 83.52X \qquad (10.1)$$

The standard errors associated with these parameters are 7.21 and 3.97 respectively, indicating that both are significant at the 5 per cent level, that the number of customers expected declines at the rate of 83.52 per mile from the Centre, and that 201.7 customers can be expected from within a mile of the Centre. (This may be a situation where it is feasible to give a physical interpretation to the value of the intercept term.)

Table 10.1 Extract of distance-related
shopping data

Distance travelled in miles (X)	Number of customers (Y)
1	199
2	161
3	109
4	75
5	66
6	46
7	39
8	12
9	22
10	5
11	10
12	7

Source: Smith (1975)

However, though it is perhaps reasonable to suggest that Y is essentially continuous, it is clear that its observed values are logically restricted to positive integers. In other words, negative values and positive real numbers are not likely to be observed in the survey data. This suggests that the classical regression approach is not appropriate. Indeed, if Y is treated as essentially categorical (in particular, a nominal variable), then classical regression is known to be inappropriate because of the violation of the constant error variance assumption (see Chapter 9).

One of the alternatives to classical regression analysis for discrete dependency relationships is the logit regression model, which was presented in detail in Chapter 9. This assumes a binomial rather than a Normal error process for the model, and uses a logit rather than an identity link function to relate the predictable mean of Y to the observed explanatory data. However, to be applicable here, the observed data would have to be supplemented by information on the number of consumers at different distances from the new shopping centre who did not patronise it. This is needed to calibrate the observed values of Y as observed proportions. As this is missing, the binomial logit model is also inappropriate.

Given the fact that the Y variable consists of simple counts rather than proportions, a second alternative model should be considered: the Poisson regression model. This model suggests that the probability of customers patronising the new shopping centre at any given distance may be calculated from the formula for the Poisson probability process (see Chapter 5):

$$p_m = \frac{\mu^m e^{-\mu}}{\mu!} \qquad (10.2)$$

where

m refers to the number of customers at given distances, and

μ refers to the mean of a Poisson distribution of customers moving a given distance

In this form, the Poisson regression model is identical to a log-linear model of the form described in Chapter 8, where the response component of the model is assumed to be the expected cell frequency.

The commands needed to specify both the classical regression and Poisson regression analyses of the data in Table 10.1 are presented in Table 10.2. The results of both analyses are presented in Printout 10.1. From the observed data (Table 10.1) we can see that the number of customers generally declines with distance from the Centre, though the pattern is rather more complex after 9 miles. The predicted values associated with the linear regression model (Printout 10.1(a)) reproduce this pattern but suggest a negative value for 12 miles. This is not a meaningful figure in this context. In comparison, the Poisson regression model also predicts a general distance decay effect. Its standardised residuals are generally smaller than the regression residuals, but as this reflects a difference in their method of calculation a direct comparison is not easy (see Chapter 8). The same applies to the deviance measures which mean different things in the two analyses. However, by calculating chi-square for both sets of data it can be shown that the Poisson model is a considerably better description of the observed data than the linear regression model.

10.2.2 Flow data: spatial interaction

The example presented in the previous subsection illustrates the use of the Poisson regression model to handle relationships which involve measurements made over distance. Extensions of this simple Poisson model may

Table 10.2 Commands to fit regression and Poisson regression models to Table 10.1

$UNITS 12	Defines length of GLIM vectors
$DATA DIST CUST	Creates two variables, DIST and CUST
$READ	Enables data entry
$YVAR CUST	Defines the response variable
$CALC D1 = %LOG(DIST)	Creates a new variable which is a natural logarithmic transformation of DIST
$ERR N	Specifies a Normal error process
$ERR P	Specifies a Poisson error process
$LINK I	Specifies an identity link for regression
$LINK LOG	Specifies a logarithmic link for Poisson regression
$FIT +D1	Command for regression analysis
$FIT +DIST	Command for Poisson regression analysis

Printout 10.1 Comparison of regression and Poisson regression analysis of Table 10.1

(a) Linear regression

$CALC D1=%LOG(DIST)
$YVAR CUST $ERR N $LINK I
$FIT D1$

CYCLE	DEVIANCE	DF
1	990.1	10

$D E$

	ESTIMATE	S.E.	PARAMETER
1	201.7	7.208	%GM
2	−83.52	3.969	D1
SCALE PARAMETER TAKEN AS			99.01

$D R$

UNIT	OBSERVED	FITTED	RESIDUAL
1	199.0	201.7	−2.689
2	161.0	143.8	17.20
3	109.0	109.9	−0.9366
4	75.00	85.91	−10.91
5	66.00	67.27	−1.274
6	46.00	52.05	−6.047
7	39.00	39.17	−0.1727
8	12.00	28.02	−16.02
9	22.00	18.18	3.816
10	5.000	9.384	−4.384
11	10.00	1.424	8.576
12	7.000	−5.843	12.84

(b) Poisson regression

$ERR P $LINK LOG$
$SIT +DIST$

CYCLE	SCALED DEVIANCE	DF
3	17.58	10

$D E$

	ESTIMATE	S.E.	PARAMETER
1	5.654	0.6088E−01	%GM
2	−0.3158	0.1427E−01	DIST
SCALE PARAMETER TAKEN AS			1.000

$D R$

UNIT	OBSERVED	FITTED	RESIDUAL
1	199	208.1	−0.6280
2	161	151.7	0.7533
3	109	110.6	−0.1558
4	75	80.68	−0.6323
5	66	58.83	0.9343
6	46	42.90	0.4729
7	39	31.29	1.379
8	12	22.81	−2.264
9	22	16.64	1.315
10	5	12.13	−2.048
11	10	8.847	0.3878
12	7	6.451	0.2161

---

be made to accommodate data in the form of flows. The term 'flow' refers to measurable movements which occur in space, for example, commuter patterns within a city, or transatlantic movements over a year. More generally, the term may be used to refer to any form of transaction which takes place between an origin and a destination. In this sense, telephone traffic between exchanges may be considered as flows even though movement is not involved.

The analysis of flow data is particularly important in geography, with many branches of the subject collecting such data. The traditional means of analysing them is to use spatial interaction models: mathematical, rather than statistical, models which are based on analytical strategies borrowed from mechanics. Excellent summaries of this form of modelling are to be found in, among others, Wilson (1974), Batty (1976), Senior (1979) and Foot (1981). Extensive use of these models has been made in various branches of geography, for example, in retailing (Huff 1963; Lakshmanan and Hansen 1965), commuting behaviour (Jensen 1980) and facility location (Massam 1975; Batty *et al.* 1985; Guy 1988).

The following are generally considered to be a 'basic set' of spatial interaction models:

1 unconstrained models (that is, models in which no assumptions are applied to either the origin or destinations);
2 singly-constrained models (either origin constrained or destination constrained in form);
3 doubly-constrained models (models in which assumptions are applied to both origins and destinations).

The unconstrained model is the simplest of the three types. In the context of retail research, where it has been used extensively, it may be written algebraically as:

$$T_{ij} = K \, O_i^{\beta} \, D_j^{\beta} \, f(c_{ij})^{-\beta} \qquad (10.3)$$

where

T_{ij} refers to the number of flows between origin i and destination j
O_i refers to some characteristic of origin i
D_j refers to some characteristic of destination j
$f(c_{ij})$ refers to a function of the travel cost between i and j
$\beta_1 \ldots 3$ are constants whose values are derived by experiment

The flows in this model may be measured in a variety of ways, for example, by calculating the number of consumers making shopping trips over a period, or by measuring the volume of expenditure. The origins and destinations may also be defined in various ways. For example, the origins may refer to a particular sub-area of a city, or to the consumers living there, whilst the destinations may refer to other sub-areas of the city, or to specified shops or shopping centres.

Given the definition of these terms, the performance of the model is determined by calibrating the constant terms empirically. This approach involves fitting a series of trial models and comparing the expected patterns of flow suggested by the model with the patterns observed in the data. Unlike statistical models, no formal mechanism exists for estimating models and testing their accuracy against chance. This has led to the suggestion that spatial interaction models suffer from the problems of over-simplicity, parameter indeterminacy, and intrinsic non-linearity (Batty 1976; Guy 1987; Davies and Guy 1987). The first of these refers to the fact that most applications of spatial interaction models tend to incorporate only one origin and one destination variable. These are general, and purport to refer to characteristics measurable at all origins and destinations. However, it is perfectly plausible that certain flows may reflect features which are unique or context-specific, for example, measurable only at a specific destination or origin. Proposals to extend the basic models to incorporate such variables have been made by, among others, Haynes and Fotheringham (1984), but Davies and Guy (1987) suggest that few of these have ever been calibrated. In spite of this, Fotheringham and Webber (1980) and Fotheringham (1983, 1984, 1985) have suggested a variety of new spatial interaction models in which extra parameters attempt to incorporate certain aspects of the spatial structure of the area concerned (for example, distance decay), or the pattern of competition between different destinations.

The second and third criticisms reflect the fact that, traditionally, the parameters of spatial interaction models have been estimated using in-efficient algorithms or numerical approximation procedures which do not ensure that the estimators are the best available. (The principal problem with any form of approximation procedure is that it provides estimates which occur at a localised maximum rather than the global maximum. The procedure within GLIM, allied to the deviance measure of fit, avoids this in most instances.) Even when apparently acceptable estimates have been produced, the lack of an explicit goodness-of-fit measure, which is also generally acceptable, has meant that it is very difficult to be sure the model is the best available. In principle, there is no reason why a model could not be developed to describe every flow correctly. However, if this is merely a saturated spatial interaction model, in the sense that it represents every flow by a corresponding parameter, then it is of little descriptive or prescriptive value. (See the discussion of saturation in log-linear modelling in Chapters 7 and 8.)

Many of these problems can be avoided if the characteristic spatial interaction model can be reformulated as a statistical model. Stetzer (1976) and Baxter (1979) have suggested various possibilities based on the use of logit or generalised odds-ratio models. However, as with the simple examples given in section 10.2.1, such approaches cannot adequately handle the presence of binary events in which all the observations are in one category. For spatial interaction models this is particularly important because a considerable number of zero flows between origins and destinations are likely to be observed in the survey data.

If the function term $f(c)$ in equation 10.3 can be expressed as:

$$f(c_{ij}) = c_{ij}^{-\delta} \qquad (10.4)$$

where

δ is a constant

then equation 10.3 can be written in a linear model:

$$\ln T_{ij} = \ln K + \beta_1 \ln O_i + \beta_2 \ln D_j - \delta \ln c_{ij} \qquad (10.5)$$

If $\ln K$ is expressed as β_0, this re-modelling of the unconstrained spatial interaction model is equivalent to a classical regression model. However, as before, because the response represents a series of counts, rather than continuities or proportions, neither the classical regression model nor the logit model is entirely suitable. A Poisson regression model is, however, suitable if it can be shown that the probability of observing m flows is given by the formula for the Poisson process:

$$p_r (T_{ij} = m) = \frac{e^{-\lambda_{ij}} \lambda_{ij}^m}{m!} \qquad (10.6)$$

where

m represents the number of flows

λ_{ij} represents the mean of a Poisson distribution of individuals moving between i and j

Flowerdew and Aitkin (1982) show that this mean value may be estimated directly from equation 10.5 in which the response is changed from $\ln T_{ij}$ to $\ln \lambda_{ij}$. By relating equation 10.6 with the characteristic expression for the Poisson distribution, the unconstrained spatial interaction model is transformed into a Poisson regression model for flows which can be specified and estimated within GLIM.

The latter factor is particularly valuable because it allows the flexibilities of the statistical modelling procedures within GLIM to be extended to spatial interaction models. Two benefits become readily apparent:

1 The criticisms levelled against spatial interaction models, which were mentioned previously, can be overcome, as GLIM provides a way of fitting a variety of different general, city-wide and context-specific variables to the same model. In addition, the estimation and fitting

strategies are robust, producing estimators and goodness-of-fit measures which are generally accepted by practitioners.

2 The gulf which has distinguished mathematical and statistical approaches in geographical teaching and research is narrowed, if not closed entirely.

(For an illustration of the two distinct approaches, see Wrigley and Bennett 1981.)

10.2.3 A retailing example of Poisson regression

Some examples of the use of Poisson regression models for the analysis of shopping data are provided in the papers by Guy (1987, 1988), and Davies and Guy (1987). The data used were taken from the Cardiff Consumer Panel Survey (Guy *et al.* 1983; Wrigley *et al.* 1985), and consist of a series of flows between fifteen residential areas (Guy 1987), or fifty consumers (Guy 1988), and eighty-three shopping centres or groups of shops within Cardiff. For the purposes of their analyses, it was assumed that the first centre visited on multi-centre shopping trips was the destination for the trip.

The commands used by Guy (1987) to fit a range of Poisson spatial interaction models are summarised in Table 10.3. The $UNITS command is set to 1,245 (15 origins by 83 destinations) and data on three variables, FLOW, DIST and SIZE, are read in on input channel 1. These variables represent:

1 FLOW: the observed flows between origins and destinations.
2 DIST: the straight-line distance between the origins and the destinations.
3 SIZE: the size of the shopping centre.

Lines 4–6 specify the response variable for the model, the probability process for *y*, and the link function. The combination for a Poisson model is selected: Poisson errors and logarithmic link. Line 7 fits a null model to the data, that is, a model which does not contain any parameters other than the grand mean. This provides an upper limit to the value of the deviance measure (see Table 10.4, line 1). Lines 8 and 9 transform the DIST and SIZE variables in logarithmic equivalents, and store the results as LDIST and LSIZE. Line 10 defines an origin-specific categorical variable (AI) containing 15 levels. These are related to the observed data values using the CALCULATE command on line 11.

Lines 12, 13 and 14 fit three alternative forms of the Poisson spatial interaction model. Line 12 fits an unconstrained model by specifying main effects terms for LSIZE and LDIST. Line 13 fits a singly-constrained model by adding the origin-specific factor to the unconstrained model. Line 14 fits an origin-specific model by fitting the main effects of the origin-specific variable, AI, and LSIZE, and the two-way interaction between AI

Table 10.3 Some GLIM commands to fit a range of
Poisson regression models (after Guy 1987)

Line number	Command string
1	$UNITS 1245
2	$DATA FLOW DIST SIZE
3	$DINPUT 1
4	$YVAR FLOW
5	$ERR P
6	$LINK LOG
7	$FIT $D E$
8	$CALC LDIST=%LOG(DIST)
9	$CALC LSIZE=%LOG(SIZE)
10	$FACTOR AI 15
11	$CALC AI=%GL (15,83)
12	$FIT LSIZE+LDIST $D E$
13	$FIT AI+LSIZE+LDIST $D E$
14	$FIT AI+LSIZE+AI.LDIST $D E$
15	$LOOK %YV %FV$
16	$STOP$

Table 10.4 Deviance measures for the models specified by Guy (1987)

Model	Deviance	DF	Change in deviance	Change in DF	Significance
1	271,900	1,244			
2	75,590	1,242	193,610	2	
3	33,670	1,238	41,920	14	
4	27,780	1,214	5,890	14	

Notes: Model 1: null model
Model 2: unconstrained model
Model 3: origin-constrained model
Model 4: origin-constrained, origin-specific model

and LDIST, representing a distance–decay effect. The results of all three models are displayed using the $D E command, and, on line 15, the observed and fitted values of the response from the final model are displayed.

The effects of these alternative models on variance reduction is summarised in Table 10.4. The null model yields a value for deviance of 271,900 for 1,244 degrees of freedom. The addition of the main effects of logged distance and logged size in the unconstrained model reduces this to 75,590 for the loss of two degrees of freedom. The addition of the production constraint, AI, reduces this still further to 33,670 for 1,238 degrees of freedom. Finally, the origin-specific model reduces deviance to 27,780 for 1,214 degrees of freedom. Parameter estimates for these models are summarised in Table 10.5, as models 2–4.

In aggregate, these singly-constrained models appear to describe the

Table 10.5 Parameter estimates of the seven models fitted by Guy (1987) to the Cardiff shopping data

Model number	LDIST	LSIZE	Parameters MULT	COOP	BIG	LVAR
2	−1.6 (436)	1.2 (271)				
3	−2.2 (352)	1.6 (249)				
4	−2.4*	1.7 (231)				
5	−1.7 (390)	0.7 (81)	0.9 (52)	0.7 (65)	0.9 (46)	0.2 (9)
6	−2.2 (330)	1.6 (127)	0.2 (8)	−0.1 (10)	0.3 (14)	−0.1 (1.4)
7	−2.4*	1.7 (121)	−0.1 (6)	−0.2 (13)	0.4 (16)	0.1 (3)

Notes: Model 2: unconstrained model
Model 3: origin-constrained model
Model 4: origin-constrained, origin-specific model
Model 5: as model 2 but including four extra terms
Model 6: as model 3 but including four extra terms
Model 7: as model 4 but including four extra terms

t values are show in brackets except for *. These parameters are estimated as the average of fifteen origin-specific estimates

flow patterns far better than the null and unconstrained models. However, when Guy compared the observed flows with those predicted by the models it was clear that even the origin-specific model seriously misrepresented some of the flows. Guy notes in particular that the model tended to overpredict trips made to the Cowbridge Road East district shopping centre, a centre lying alongside a major arterial routeway to the west of Cardiff city centre, whilst underpredicting flows to the city centre, a district centre in North Cardiff, and small centres located close to residential zones.

In order to improve the description of more of these flows, Guy calibrated a further series of three models incorporating additional explanatory information on the facilities available in each shopping centre. The purpose of adding these variables was to improve the specification of the attractiveness terms in the model. These were:

1 MULT: a binary dummy representing the presence or absence of a multiple grocery store (typically, one owned by a local or national chain).
2 COOP: a binary dummy representing the presence or absence of a co-operative grocer.

3 BIG: a binary dummy representing the presence or absence of a large grocery outlet (exceeding 1,500 square metres' sales area).
4 LVAR: a logged variable used to represent the number of different types of shop to be found in the centre.

The effects of these on deviance reduction are summarised in Table 10.6. Guy notes that these models appear to reduce unexplained variation significantly in aggregate, whilst only providing a limited improvement in individual flow prediction. The flows to the city centre were described better using these extended production constrained models, but flows to small centres near residential areas were still erratic.

Table 10.6 Deviance measures for the extended models specified by Guy (1987)

Model	Deviance	DF	Change in deviance	Change in DF
1	271,900	1,244		
5	66,280	1,238	205,620	6
6	33,250	1,224	33,030	14
7	26,850	1,210	6,400	14

Notes: Model 1: null model
Model 5: unconstrained model (model 2) plus four extra variables
Model 6: origin-constrained model (model 3) plus four extra variables
Model 7: origin-constrained, origin-specific model (model 4) plus four extra variables

These examples illustrate the basic approach to spatial interaction modelling using generalised linear models. They have shown that it is relatively easy to specify and test a variety of different types of interaction model, yielding measures for aggregate fit which are significant. However, Guy notes that the prediction of individual flows tends to be poor. Indeed, even the incorporation of effects to represent competing destinations (as suggested in Fotheringham 1983, 1985) does not produce a marked improvement in performance. Ironically, Guy's results are at variance with those obtained by Fotheringham, suggesting the need for further research in this area.

10.2.4 A historical example of Poisson regression

A second example of the use of the Poisson regression model for analysing flow data comes from a paper on migration fields by Lovett *et al.* (1985). This paper is concerned with illustrating some of the conditions affecting in-migration to a pre-industrial city: Edinburgh in the seventeenth and eighteenth centuries.

Historical research appears to indicate that in-migration was a major factor in maintaining the sizes of many pre-industrial towns and cities at times when levels of urban mortality were high (Patten 1976; Agnew and Cox 1980; Holman 1980). However, in spite of this, Lovett *et al.* suggest that relatively little is known about the factors which generate and shape in-migration flows at this time. In particular, they note that many existing studies:

1 appear only to focus on individual towns or cities for short periods of time, thus failing to capture the essentially dynamic nature of migration;
2 have tended to examine the source areas for the migration streams at coarse or generalised spatial scales;
3 have failed to model sufficiently the socio-economic and cultural factors which are known to be involved (for a detailed examination of these factors in the case of Gaelic speakers in Scotland, see Withers 1981, 1984).

Indeed, many studies merely present the flows between areas as percentages of the total amount of movement in a specified period, thus ignoring their essentially nominal character.

Poisson regression was considered to be a viable descriptive tool in this context because it provides a means of handling counts, many of which are likely to be very small or zero, whilst also linking the observed patterns of movement with possible explanatory variables. The data for the study came from four sources:

1 The apprenticeship records for Edinburgh published by the Scottish Record Society provided the observed flows (they contain information on the name of the apprentice, his father's name, his father's occupation and domicile, the name and trade of the master, and the date of entry).
2 Hearth tax returns for 1691 (population data for the late seventeenth century).
3 Webster's Census for 1755 based on parish clergy (population data for mid-eighteenth century).
4 1801 Population Census (population information for late eighteenth century).

By linking the information in these together, Lovett *et al.* were able to analyse movements out of thirty-three counties and eighty-four burghs for three temporal cross-sections: 1675–99, 1725–49, 1775–99. (For further details, and a discussion of some of the problems of these data sets, see the 'Data and Hypotheses' section of their paper.)

The initial analysis of the county-based data focused on regressing the number of flows against the natural logarithms of population size and distance from Edinburgh. The results of this are presented in Table 10.7. Though the overall fits of the model are good, the structure of the residuals indicated that further explanatory information could be included. After some experimentation, they included the following:

1 A measure of urbanisation in each area of origin.
2 A categorical measure which they term a sectoral classification, to represent the orientation of each county with respect to Edinburgh.
3 A series of three origin-specific variables which allow the effects of distance, urbanisation and population variables to be assessed by sector.

Table 10.7 Deviance measures for the initial analyses of Edinburgh in-migration

Model	Deviance	DF	Change in deviance	Change in DF
1675–99				
null	1,154.0	17		
+LD+LP	201.6	15	952.4	2
1725–49				
null	938.8	32		
+LD+LP	103.8	30	835.0	2
1775–99				
null	1,350.0	32		
+LD+LP	159.2	30	1,190.8	2

Notes: LD = natural logarithm of distance
LP = natural logarithm of population

The number of observations is reduced to 18 for the 1675–99 cross-section because of a lack of data for the Highlands and the south-west of Scotland

The results of these extended models are presented in Table 10.8. The number of observations is reduced to eighteen for the 1675–99 cross-section because of a lack of data for the Highlands and the south-west of Scotland.

Based on these findings, and supplemented by the analysis of burghs, Lovett *et al.* concluded that in each time cross-section, origins in the southern sector (that is, in the origins located near the English border) had the highest levels of apprenticeship out-migration to Edinburgh relative to their population, distance and urbanisation conditions, whilst, in the west, the opposite held. They also noted that the migration field of the city contracted with time as alternative competing and intervening opportunities (for example, Glasgow, Tayside counties) developed. This contraction was not uniform either spatially or through time. They concluded:

Distance from the capital and the population of centres of origin were important variables in explaining levels of migration but the city's migration field contained clearly-defined flows from particular areas even after the effects of distance and population had been allowed for.

(Lovett *et al.* 1985: 330)

Table 10.8 Deviance measures for the extended analyses of Edinburgh in-migration

Model	Deviance	DF	Change in deviance	Change in DF
1675–99				
null	1,154.0	17		
+LD+LP	201.6	15	952.4	2
+U	110.5	14	91.1	1
+S	37.1	12	73.4	2
+S.LP	28.3	10	8.8	2
+S.U	23.3	8	5.0	2
+S.LD	12.3	6	11.0	2
1725–49				
null	938.8	32		
+LD+LP	103.8	30	835.0	2
+U	91.9	29	11.9	1
+S	59.4	27	37.0	2
+S.LD	44.2	25	15.2	2
+S.LP	35.9	23	8.3	2
+S.U	22.6	21	13.3	2
1775–99				
null	1,350.0	32		
+LD+LP	159.2	30	1,190.8	2
+U	91.9	29	67.3	1
+S	80.5	27	11.4	2
+S.LD	51.8	25	28.7	2
+S.LP	43.1	23	8.7	2
+S.U	32.6	21	10.5	2

Notes: LD = natural logarithm of distance
LP = natural logarithm of population
U = urbanisation variable
S = sectoral variable

The number of observations is reduced to 18 for the 1675–99 cross-section because of a lack of data for the Highlands and the south-west of Scotland

They also noted that many of these flows persisted through time, but were influenced by evolutional changes in the migration field as alternative destinations emerged, or as economic prospects in many former areas of origin made out-migration less necessary.

10.2.5 Quasi-likelihood and pseudo-likelihood models

The two examples given in the last two subsections show that a statistical treatment of flow data is possible, providing researchers with a flexible modelling and inferential strategy. However, before these inferences may be considered valid, it is necessary to ensure that the assumptions of the Poisson regression model are relevant to the social or environmental context being modelled (for an account of these assumptions, see Chapter

5, section 5.5). One simple example illustrates the general point. In a Poisson distribution, the mean and variance values are identical. If, however, the flows in a model designed to describe shopping behaviour were measured as expenditures, then these are clearly:

> not counts of independent outcomes as would be appropriate for a Poisson model. Indeed, far from having an expected value of unity, the mean/variance ratio would depend upon the arbitrary decision of what monetary units to use in measuring expenditure.
>
> (Davies and Guy 1987)

In their paper, Davies and Guy discuss various ways of generalising the Poisson distribution to account for anticipated departures from its underlying assumptions. Two departures are most likely:

1 *extra-variation:* a departure which occurs because relevant explanatory variables have been omitted from the linear predictor (see also Kennedy 1979: Chapter 5);
2 *temporal dependence*: a departure which arises because shopping trips in any given period are likely to be interrelated.

The latter reflects consumer behaviour and will vary from product to product. However, it seems fair to assume that for many types of product, particularly those described by shopping models, consumption will decrease if stocks are already high. The numbers and types of shopping flows experienced at the end of a monitoring period will thus not be independent of those preceding them. Both may be accommodated by generalising the Poisson, producing via two alternative routes the negative binomial distribution (see Chapter 5, section 5.7.1).

The specification and estimation of a negative binomial spatial interaction model is relatively easy in GLIM if the model being applied is unconstrained. Unfortunately, the application of constraints, either to the origin, destination, or both, considerably increases complexity if the model is not Poisson, because the constraints must be modelled explicitly. Davies and Guy note that this modification is too complex for specification within GLIM or other commercial packages, and would require researchers to resort to special-purpose software such as provided by the NAG or Harwell subroutine libraries. Moreover, as the negative binomial model does not ensure consistent parameter estimation if the data are not negative binomial in the first place, it is unlikely that even this procedure would resolve the problem satisfactorily.

The approaches they suggest as suitable alternatives are to estimate Poisson regression models under quasi-likelihood and pseudo-likelihood conditions. The quasi-likelihood approach was introduced in Chapter 6 to illustrate how robust estimators could be generated from the first two moments of a distribution. Thus, instead of assuming full knowledge of the shape of a sampling distribution (as is required in full likelihood estimation), the researchers can simply employ their more limited knowledge to yield estimators which are at least as good as those obtained under the full model. Indeed, as was described in Chapter 6, the maximum likelihood parameter estimates and the maximum quasi-likelihood parameter estimates are identical. However, the standard errors of parameters are different, though related.

Pseudo-likelihood differs from conventional likelihood and quasi-likelihood in that it does not impose the restriction that the underlying model being specified is correct (White 1982; Gourieroux *et al*. 1984). This means that it can provide for the estimation of a wider range of misspecified Poisson regression models than either of the alternatives. The characteristics of this approach are still in development and there are comparatively few empirical examples of its use in geography. However, it can be shown from theory that the consistency of the parameters of a Poisson regression model only requires that the mean is correct. Nothing further is required in order to generate acceptable point estimates of population parameters. As a result, the parameter estimates from full likelihood, quasi-likelihood and pseudo-likelihood Poisson regression models will be the same. However, the standard errors of these parameters will be different because these are sensitive to misspecifications. Baxter (1983) and Davies (1987) suggest some ways of calculating robust estimates of these standard errors, and indicate that, in general, standard errors from misspecified full-likelihood and quasi-likelihood models will be potentially unreliable.

10.2.6 Summary

The advantages of the Poisson models of spatial interaction may be summarised as follows:

1 They provide greater flexibility in model specification than is generally available using traditional models or percentage flows.
2 They allow standard diagnostic tests for model specification and assessment to be applied.
3 They provide a mechanism for extending the basic set of origin and destination variables allowing context-specific variables to be included in the model.
4 By being specified in a form capable of estimation in GLIM, they benefit from the efficiency of the program's maximum likelihood algorithm.

These advantages significantly assist researchers interested in the analysis of flow data who previously were forced to use *ad hoc*, special-purpose software, which frequently failed to provide fully acceptable estimates of model parameters. In addition, by providing a statistical treatment of an area of quantitative research traditionally the preserve of mathematical models, the Poisson regression generalised linear model offers a method of integrating two areas of activity which have frequently been considered to be distinct, if not independent.

However, though a useful advance on previous experience, the computational complexity of estimating misspecified constrained spatial interaction models still remains a major obstacle. Research is still needed to investigate further the areas of pseudo-likelihood estimation, robust estimation in the face of misspecification, and model generalisation.

10.3 EXPERIMENTAL DESIGNS AND DESIGNED EXPERIMENTS

Many of the most significant developments in the theory of statistics have been associated with work on designed experiments. The object of experimental designs is to provide a testing ground for complex hypotheses which is as controlled as possible. There are many such designs which differ in their complexity and the degree of control they aim to apply. Though their use is relatively uncommon in geography, particularly in human geography, interest does appear to be increasing as a result of their application to models of consumer choice. A number of recent articles have incorporated some form of experimental design in their analysis, for example, Timmermans (1980), Wrigley (1980), Longley and Wrigley (1984), Bates (1986) and O'Brien (1987a).

10.3.1 Some terminology

In any experimental design attention focuses on how some subject responds to a variety of alternative stimuli or treatments, one of which may be a control or placebo. The object of the design is to see if the response pattern is random or systematic, and if the latter, to identify possible reasons for it. By organising the nature of the experiment, researchers aim to eliminate, or otherwise standardise, the many factors other than the stimulus which may influence the response pattern.

The simplest experimental design is the completely randomised design (Chatfield 1983) in which subjects are allocated to experimental groups completely at random. An example of this is the double blind experiment used in clinical trials of a drug or a treatment, in which neither the subjects nor the experimenters who are in day-to-day charge of the experiment know who is receiving what. More complex designs can be produced by blocking the subjects into discrete, non-overlapping experimental blocks in which the treatments are allocated at random. Examples of this are the Latin Square, Graeco-Square and their variants. (For an illustration of the use of a Latin Square in assessing traffic flow at various sites in a city over a seven-day period, see Haggett 1975.)

Printout 10.2 Transcription of Cormack's analysis of capture–recapture data

```
£r *glim
$UNITS 32
$DATA N
$READ
1 0 5 1 0 1 2 5 0 4 2 7 4 2 1 19
1 1 0 3 0 2 0 9 9 4 1 13 10 13 21 100
$CALC A=%GL(2,1)−1:B=%GL(2,2)−1:C=%GL(2,4)−1
$CALC D=%GL(2,8)−1:E=%GL(2,16)−1
$CALC W=1
$CALC AB=A*B:BC=B*C:CD=C*D:DE=D*E
$CALC B3=AB*C:B4=B3*D:D2=DE*C:D1=D2*B
$CALC PBD=B+C+D:PB=PBD+E:PD=PBD+A
$EDIT 32 W 0
$LOOK N A B C D E$
    —— LIST TRUNCATED
```

	N	A	B	C	D	E
1	1.000	0.0	0.0	0.0	0.0	0.0
2	0.0	1.000	0.0	0.0	0.0	0.0
3	5.000	0.0	1.000	0.0	0.0	0.0
4	1.000	1.000	1.000	0.0	0.0	0.0
5	0.0	0.0	0.0	1.000	0.0	0.0
6	1.000	1.000	0.0	1.000	0.0	0.0
7	2.000	0.0	1.000	1.000	0.0	0.0
8	5.000	1.000	1.000	1.000	0.0	0.0
9	0.0	0.0	0.0	0.0	1.000	0.0
10	4.000	1.000	0.0	0.0	1.000	0.0
11	2.000	0.0	1.000	0.0	1.000	0.0
12	7.000	1.000	1.000	0.0	1.000	0.0

```
$YVAR N
$ERR P
$WEIGHT W

$FITS
```

	SCALED	
CYCLE	DEVIANCE	DF
5	181.5	30

`$FIT A+B+C+D+E$`

	SCALED	
CYCLE	DEVIANCE	DF
4	67.27	25

`$FIT +AB+B3+B4$`

	SCALED	
CYCLE	DEVIANCE	DF
5	56.83	22

`$FIT +DE+D1+D2$`

	SCALED	
CYCLE	DEVIANCE	DF
4	37.07	19

`$FIT +AB+BC+CD+DE$`

	SCALED	
CYCLE	DEVIANCE	DF
4	36.88	17

```
$D ER$
```

	ESTIMATE	S.E.	PARAMETER
1	0.7109E−01	0.4111	%GM
2	0.7696E−01	0.2775	A
3	−0.1839	0.4411	B
4	−0.2076	0.5843	C
5	0.8060	0.4408	D
6	0.6454E−01	0.3592	E
7	1.022	0.4924	AB
8	1.317	0.8284	B3
9	−0.3109E−01	0.6831	B4
10	0.6663	0.4927	DE
11	1.008	0.7315	D1
12	0.7464	0.5726	D2
13	−0.1824	0.7860	BC
14	0.2564	0.7088	CD

SCALE PARAMETER TAKEN AS 1.000

UNIT	OBSERVED	FITTED	RESIDUAL
1	1	1.074	−0.7111E−01
2	0	1.160	−1.077
3	5	0.8933	4.345
4	1	2.680	−1.026
5	0	0.8724	−0.9340
6	1	0.9422	0.5959E−01
7	2	0.6048	1.794
8	5	6.774	−0.6817
9	0	2.404	−1.550
10	4	2.596	0.8713
11	2	2.000	−0.6651E−04
12	7	6.000	0.4082
13	4	2.524	0.9290
14	2	2.726	−0.4397
15	1	1.750	−0.5669
16	19	19.00	0.4157E−05
17	1	1.145	−0.1357
18	1	1.237	−0.2130
19	0	0.9529	−0.9762
20	3	2.859	0.8365E−01
21	0	0.9305	−0.9646
22	2	1.005	0.9926
23	0	0.6452	−0.8032
24	9	7.226	0.6600
25	9	4.993	1.793
26	4	5.392	−0.5995
27	1	4.154	−1.548
28	13	12.46	0.1525
29	10	11.06	−0.3181
30	13	11.94	0.3061
31	21	21.00	0.4370E−05
32	100	228.0	0.0

$STOP$

Note: Commands prefixed by £ are operating system commands
Commands prefixed by $ are GLIM commands

More complex than these blocked designs are factorial designs in which several different types of treatment, rather than alternatives of a single treatment, are analysed simultaneously. These designs may be either balanced or unbalanced, depending on whether every possible treatment type is applied to each subject. For a detailed account of the literature and methodology of experimental designs see, among others, Cochran and Cox (1957), Anderson and Bancroft (1952), John (1971) and Namboodiri *et al.* (1975).

Many of the ideas present within the experimental design literature may be manipulated to accommodate research problems in the social and environmental sciences. Some of these procedures may be of value to the geographer. One design which is potentially valuable, particularly to biogeographers, accommodates the Jolly–Seber models of birth and death for capture–recapture experiments in environmental populations.

10.3.2 Capture–recapture experiments

Capture–recapture experiments are frequently used in studies of biological populations when an assessment of their size or characteristics is required. The methodology involves capturing samples of animals at two time periods and comparing their characteristics. The differences between the two samples may be used to assess characteristics in the whole population. Many of the procedures which are used in these experiments are based on generalised linear models and so may be fitted to experimental data using GLIM. For a discussion of the mathematics of capture–recapture experiments, see Bishop *et al.* (1975).

Fienberg (1972) and Cormack (1981) show that it is possible to use log-linear models to analyse capture–recapture experiments applied to both open and closed populations, incorporating parameters to reflect the effects of births, deaths and migration. An example of how GLIM may be used for this type of analysis is given by Cormack (1980). The raw data are taken from a study by Manly and Parr (1968) and consist of the complete capture histories of all animals seen during a time interval, recorded as a 2^5 factorial design. A transcript of part of Cormack's analysis is presented in Printout 10.2.

The models applied to the Manly–Parr data are set up in the standard way for log-linear models, with \$YVAR being set to the observed cell count, and a Poisson process–logarithmic link function being selected for the expected cell frequencies. However, a number of differences exist because the biological nature of the experiment prohibits the fitting of hierarchical effects. The main effect of this is that the variables are not defined as factors, but instead are generated as a series of binary vectors.

In analysing these data, a closed population is represented by a log-linear model containing only the main effects terms (%GM+A+B+C+D+E), whereas death and birth between two time periods are represented by appropriate two-way interactions. The models for a closed population, for birth and death, and for 'trap dependence' are summarised in Table 10.9. On the basis of the deviance statistics, Cormack suggests that birth and death effects are both significant, but that trap dependence between successive periods is not. By looking in more detail at the standardised residuals from this model, further reductions in deviance are obtained, encouraging Cormack to suggest that temporary emigration of one of the species in the study had taken place between the first and second sampling periods.

Table 10.9 Analysis of deviance for Manly–Parr capture–recapture data

Model	Deviance	DF	Change in deviance	Change in DF
%GM	181.5	30		
Closed population	67.3	25	114.2	5
Birth	56.8	22	10.5	3
Death	37.1	19	19.7	3
Trap dependence	36.9	17	0.2	2

10.4 DISCRETE CHOICE MODELS

10.4.1 Choice and preference

The analysis of choice and preference behaviour is commonplace in several social science disciplines. In economics, for example, the notion of utility is used to determine why a rational consumer chooses one product rather than another. Psychology also makes use of a similar idea in its study of preference behaviour. An individual is assumed to gain a greater degree of preference or benefit by choosing one state rather than another. In biology and pharmacology, the notion of a tolerance distribution is used to interpret results in which objects tend towards one state rather than another.

Utility, preference and tolerance are implied in each of these situations as being properties or characteristics which goods or other phenomena may possess. At no time is a measurable object or phenomenon described. This being the case, it is important to realise that the behaviour of the subject is interpreted as if utility (preference, tolerance) actually existed. There is a clear circularity here between the observed response and the assumed cause of that response: a numerical measure may describe some empirical happening, the unobserved, but assumed, utility (preference, tolerance distribution) provides for its explanation.

10.4.2 Extensive and intensive margins

The model of choice which underlies the development of discrete choice models distinguishes between choices which are made at the extensive margin and those at the intensive margin. The reason for this distinction is that many commodities can only be chosen if they are consumed in their entirety as discrete 'bundles'. For example, in the 'choice' of a mode of transport to get to work (school, shops, etc.) the alternatives available to the subject may be:

1 walk;
2 take the bus;
3 take a taxi;
4 go by car as a driver;
5 go by car as a passenger.

Though it is perfectly possible to make use of more than one of these modes in the same journey (for example, walk to the station, take a train, change on to the underground, take a taxi), each stage is accomplished by a single mode to the complete exclusion of the other alternatives. This sort of choice problem corresponds to the extensive margin, and is seen to be different from that facing a shopper who might purchase 1/2 lb, 1lb, or more than 1lb of butter. In this case, the commodity being selected, butter, is divisible; it may be consumed in varying amounts and not just as an 'all-or-nothing' decision. Hensher and Johnson (1981: 12–13) note that the definition of the extensive margin is rather more complex than this, in that an individual may only consume a divisible commodity in bundles, hence perceiving it as discrete rather than continuous. For many social science problems it is perhaps feasible and reasonable to consider behaviour in terms of these 'all-or-nothing' decisions.

10.4.3 Some algebraic models of discrete choice

Theoretical frameworks of individual choice based on a concept of random utility maximisation have been developed by, among others, Arrow (1951), Luce (1959), Lancaster (1966, 1971) and Rosen (1974). Many of these may be expressed as categorical regression models using the logit transformation (see McFadden 1974; Richards and Ben-Akiva 1975; Hensher and Johnson 1981; Ben-Akiva and Lerman 1985; Wrigley 1985; and Train 1986 for further details). In this format, the model imposes highly restrictive assumptions on decision processes and has been found to be quite inappropriate for many apparent choice problems. Computationally tractable models capable of representing less restrictive formulations of discrete choice may be based on the probit transformation (Daganzo *et al.* 1977; Hausman and Wise 1978; Daganzo 1979; Manski and

McFadden 1981), or alternatively, on a series of 'half-way houses' lying somewhere between logit and probit:

1 nested (structured or hierarchical) logit (Sobel 1980);
2 dogit (Gaudry and Degenais 1979; Gaudry 1980);
3 generalised extreme value logit (Manski 1981);
4 cross-correlated logit (Williams and Ortuzar 1982);
5 weight-shifting models (Meyer and Eagle 1981).

It is important to remember that when developed in the framework of random utility maximisation, these choice models represent particularly specialised versions of their statistical counterparts which were described in Chapter 9.

GLIM may be used to fit a number of these discrete choice models. However, for the avid user, specialised packages such as BLOGIT and QUAIL are likely to be of more value. This is because they can cope with multi-way choice problems, whereas GLIM is essentially limited to binary choice (except through its use of special macros), and are more amenable to the specification of complex decision rules.

10.4.4 Some fundamental difficulties

Apart from these computational difficulties, there are a number of major conceptual problems with the use of random utility maximisation discrete choice models. These may be divided neatly into areas of substantive difficulty within the accepted discrete choice methodology, and areas of conflict with this methodology. The former are in many ways technical issues which may be amenable to a technical solution, for example, the selection of the members of the choice set, classification, sampling and measurement problems. However, the latter strike at the heart of the current discrete choice methodology, in that they question the behavioural postulates underlying the use of the models. For further details of these issues, see Deaton and Muellbauer (1980) and Blundell (1988).

10.5 CONCLUSIONS

The aim of this chapter has been to show how some of the statistical models now available to assist geographers tackle numerical problems involving mixtures of categorical and continuous data may be extended or developed into highly specialised uses. It is not implied that all of these methods will be relevant to all geographers all of the time, but there may be situations where some geographers may find some of these methods helpful.

Methods such as Poisson regression are simply extensions of common linear regression ideas, which are found to be useful under certain distributional conditions. The use of such models does not imply a major commitment to a particular style of modelling or view of the world. Specialist developments such as experimental design models and discrete choice models are very different. In both approaches a considerable body of philosophical and theoretical argument is invoked, even if the invocation is implicit. This is certainly the case in the context of the designed experiment which has fostered a particular view of scientific practice that is still widely and usefully held. Discrete choice models, on the other hand, represent an explicit attachment to a particular view of individual decision-making. This view, given appropriate modification to make naïve models more 'realistic', supports a variety of mathematical formulations of random utility maximisation. However, the very success of this view in certain areas of social research and engineering should not obscure the fact that views diametrically opposed to random utility maximisation are held by many practitioners interested in the study of decision-making.

11

SUMMARY AND CONCLUSIONS

11.1 INTRODUCTION

The need to reassess quantitative geography has become only too apparent in recent years as the range of techniques and strategies to be covered has grown. The days of courses consisting mainly of descriptive measures, graphical devices, a number of non-parametric techniques for under-measured data, and culminating in correlation and regression have long past. New avenues of work have been established which have significant bearing on quantitative teaching. One of these is the modification of existing statistical techniques and procedures to spatial data. A second is the incorporation of categorical models given that geography is over-flowing with such data. A third is the need to introduce the idea of robustness so that students are sensitive to the fragility of many statistical techniques. A fourth is the need to computerise quantitative training so that students have the technical apparatus and experience to work with the large data holdings stored in geographical information systems and continent-wide data bases.

In order to keep abreast of these developments quantitative geography needs to be organised so that stress is placed on the development of a consistent approach to data handling rather than the documentation of a diverse list of techniques. Given the nature of the material, the latent hostility of students towards numbers, and the shortages of time, technical assistance and teaching support, this is frequently not possible unless topics are left out or a framework can be developed to provide a clearer focus. Nelder and Wedderburn's class of generalised linear models provides the latter.

This book has attempted to provide an introduction to this approach by presenting some of the key techniques in current use in generalised form. It has also attempted to link quantitative modelling back to data description because models are merely one way of summarising numerical data rather than some sort of pinnacle of quantitative achievement. However, generalised linear models do not solve all quantitative problems for the geographer. They are presented here as a way of making life easier for all

313

concerned with quantitative geography; they are not a panacea for geographical problems.

This chapter tries to draw some of the main themes of the book together by looking at three types of integration provided by generalised linear models and three types of limitation.

11.2 HORIZONTAL INTEGRATION

The main theme to be presented throughout this book is that many of the most important linear statistical models in use in geography are members of a common class. In Part II an attempt was made to show how the class of generalised linear models could be used to specify five major linear statistical models: the linear regression model for Normally-distributed continuous data; the analysis of variance model for Normally-distributed categorised data; the log-linear model for contingency table data; and the logit and probit models for proportions data. Each of these models has been described in the quantitative literature of different academic disciplines for many years, sometimes being reinvented afresh several times. This has resulted in the generation of a body of quantitative literature which is highly repetitive and sometimes conflicting in the advice it provides. The most significant drawback, however, is that versions of the same model appear in several disciplines but the fact that the models are identical is lost behind numerous differences in notation and terminology. In nearly all cases these differences are superficial and obscure the learning process.

The problem for the geographer, student and quantitative methods teacher alike, is that an increasing number of geographical research papers have been published since the mid-1970s which make use of all of these models. Some of the categorical models have also become core analytical tools in certain areas of geography, for example, in the study of transport and intra-urban mobility. In these areas, these are the normal tools of research rather than the exceptions.

A popular approach which attempts to link all five models is the 'didactic tableau' (Figure 11.1). In this design the traditional models of quantitative geography courses – linear regression, dummy variable linear regression and the analysis of variance – are displayed along the first row as cells (a), (b) and (c) respectively. These three models are distinguished by the form of their explanatory variables, i.e., though each of these models possesses a continuously-distributed response variable, they differ in the character of their explanatory variables. Cell (a) is described by continuous explanatory variables, cell (c) by categorical explanatory variables, and cell (b) by a mixture of both. This idea, applied to the second row, provides a way of linking the categorical models (generally unfamiliar to most geographers) with the traditional models. Thus the probit model is allocated to cells (d)

314

Explanatory variables

		Continuous scale	Mixed scales	Categorical scale	None specified
Response variable	Categorical scale	a	b	c	
	Continuous scales	d	f	e	g

Figure 11.1 Didactic tableau
Source: Adapted from O'Brien and Wrigley (1980).

and (e), the logit model to cells (d), (e) and (f), and the log-linear model to cells (f) and (g). Cell (f) represents a 'zone of transition' in that log-linear models, suitably re-expressed, and logit models may both be used to tackle asymmetric contingency table problems. The reason for this is that the log-linear model may be reformulated to mirror the performance of the logit model (Haberman 1974a; Bishop *et al.* 1975).

The principal advantage of this type of display is that it builds on information on regression and the analysis of variance which many geographers should already have acquired from introductory quantitative methods courses. Second, the distinctions between response and explanatory variables and categorical and continuous scales of measurement seem simple and straightforward. Unfortunately, these 'strengths' are also potential weaknesses because:

1 many geographers do not have a sufficient grasp of linear regression and the analysis of variance to use them as building blocks;
2 the distinction between response and explanatory variables is often not easy to make and, indeed, may be artificial (Plackett 1974);
3 no attempt is made in the tableau to find similarities between the five models other than at a surface level; and
4 the models are still presented as five distinct and separate topics, each with its own notation, intellectual heritage, and *modus operandi.*

In effect, the use of the tableau to present these five models may generate considerable additional work without any real increase in knowledge being acquired. Moreover, if the preparatory work on regression has not been established in an introductory course, the whole design becomes unintelligible.

Nelder and Wedderburn's class of generalised linear models offers an alternative framework which should make life easier for all concerned. Though not developed specifically to provide a framework for teaching, its very design makes it particularly suitable for this task. The core equation of the class:

315

$$y_i = g^{-1} (\Sigma \beta_k X_{ik}) + \varepsilon_i \qquad\qquad (11.1)$$

provides a structure which is simple to learn, and which can be considerably modified to represent many types of dependency and inter-
dependency relationship, and many different types of measurement. In this
equation the nature of the components are not determined prior to
analysis. If the problem calls for it X may be treated as an independent
variable matrix composed of fixed components and β as a vector of
Normally-distributed random errors. Set like this, the equation represents
the standard format of a least squares linear regression model. Alternatively,
X may be treated as a design matrix composed of ones or zeros, producing
a factorial structure typical of the log-linear model. In this context, ε is not
defined explicitly, nor are distributional assumptions normally applied to
it. Between these two formulations of X (essentially equivalent to specifying the problems of cells (a) and (g)) lies a range of alternatives which
allow the remaining models in Figure 11.1 to be specified.

The key advantage of this form of presentation is that categorical and
continuous statistical models may be specified within the confines of a
single equation. As a result historical differences in notation, derivation,
mode of inference and estimation may be ignored as the generalised core
relationship provides the necessary tools to specify each. The result
therefore is a considerable horizontal integration of the five models which
offers the geographer a better basis for handling the diversities of geographical data sets.

11.3 VERTICAL INTEGRATION

In addition to horizontal integration, the class of generalised linear models
may be integrated vertically. This means that the experience of modelling
may be linked backwards towards the simple descriptive summary
measures which are key components of introductory quantitative methods
courses. One of the most frequent drawbacks to quantitative teaching in
geography is that different people are responsible for the teaching of
introductory courses and the more advanced modelling courses. As a
result, there is often little direct overlap between the two experiences and
geographers find themselves having to bridge an intellectual gap between
description and modelling.

The main theme of Part I was to show how models could be introduced
as extensions of simple descriptive techniques. These were shown to be of
value if the relationships in the data are extremely straightforward (e.g.,
the six-times table) or are difficult to describe. Skew and multimodal data
are obvious examples of 'difficult' data types which are frequently met in
geography. In analysing these, attention must be paid to the form of the
data set, to its shape and the relative magnitude of its observations, as

these may severely affect the performance of measures. Simple summary measures such as the mean, median, standard deviation and variance are all affected by such data. Robust measures provide some help, at least in identifying the severity of the confounding. However, they do not provide a fully complete answer. Models may be able to do better.

Models and summary measures may thus be presented as merely two positions along the same dimension of data description. Knowledge of this fact may help geographers to see how the two topics fit together. Another feature which should be noted is that most of the models which are typically used by geographers are based on the comparison of mean–variance relationships. Attention is paid in particular to comparing variability within specific variables with that between variables. This idea was illustrated by all the examples in Part II, but was introduced originally in the extended examination of the data in Table 4.1. A preliminary analysis of data (not model-based) also brings out the need to look at variability between and within variables.

There is a third and rather more technical sense in which generalised linear models and data description may be integrated vertically. This concerns the ability to produce estimates of the parameters of generalised models. In the traditional descriptions of these models assumptions are usually made about an underlying probability process (e.g., normality for the regression model, Poisson for log-linear models, etc.) in order to obtain acceptable estimates of their parameters. One of the main features of Nelder and Wedderburn's work which allows the class to be developed is the fact that the key properties of its parameters depend less on the assumption of a specific underlying probability distribution than on the form of the mean–variance relationship and on a lack of correlation in the error component. Both of these topics are usually presented in introductory courses and it is not difficult to see that they may be used to bring the realms of simple description and modelling closer together. In other words, when students are being instructed in handling data particular attention is placed on mean–variance relationships through the use of numerical summaries and graphics. When attention moves on to consider models, the key features which still need to be studied are mean–variance relationships and the character of the error component. The link between the two topics is thus very clear.

11.4 COMPUTATIONAL INTEGRATION

The final point presented in the previous section is also relevant here because it provides the focus for the computational integration provided by generalised linear models. In a traditional quantitative geography course a major factor limiting what can be done is the availability of computer software. The choice is further complicated by the need to ensure that relatively little time will be spent learning to compute or, if the software is commercial, make the package work. A much more valuable experience is gained if the bulk of the computing time is spent applying models to data and assessing their performance.

There are many commercial packages on the market which can support quantitative geography. Many of these are organised in modules, each of which corresponds to a specific technique. In this design the user calls up whichever module is required to perform a specific type of analysis. The advantage of this is that only those modules which are to be used need to be considered in detail. Problems can occur, however, if several modules are to be used which have different command languages, as the task of learning the idiosyncracies of individual modules is clearly inefficient. More seriously, if the data cannot easily be swapped between different modules, comparative analysis is made very difficult. The problem is compounded if the analytical module required for comparison is not part of the package and there are no facilities to interface home-made software.

These problems are encountered with many of the popular social science packages in everyday use in the UK and USA. Packages such as MINITAB, SPSSx and BMDP provide a variable range of modelling options, but are very different in their ease of use. This leads to a common (and unhealthy) reaction: the tendency for many geographers to stick rigidly to a package they have 'cracked'. Unfortunately, though this may minimise the problems mentioned previously, it means that they limit themselves only to conducting analyses which are supported by their favoured package.

This reaction illustrates the tension which exists between the need to provide a computing environment which is easy to learn and one which is flexible, allowing geographers to experiment with standard analytical alternatives and ones they have devised for themselves. A user-friendly package containing standard techniques may be of teaching value but of limited research value. On the other hand, a flexible computing environment which allows geographers to do virtually what they like may be little better than using a high-level computer language. For many geographers this alternative is a step in the wrong direction.

Generalised linear models offer a solution to this computing problem which provides flexibility and a common command language. They can do this because maximum likelihood estimates of the parameters of any generalised linear model may be obtained using a single algorithm. There is thus no need to organise a computer package into distinct, unrelated, modules, or to develop a complex menu-driven command language to specify options. GLIM represents the most comprehensive attempt to provide a computationally integrated package for fitting members of the class. In this, a series of simple commands need to be specified before the package may fit models to data. The combination of a small number of commands determines which model from the class is fitted. By developing the package in this way users are offered the greatest modelling flexibility possible. Once the data have been read in, any suitable member of the class may be fitted merely by selecting a different combination of commands.

Details of model fitting in GLIM are presented throughout Part II and in Appendices A and C. In addition, it is also worth knowing that modifications have been made to GLIM to allow it to handle data description, the theme of Part I. These are not as advanced as those in packages such as MINITAB, but the development suggests that NAG is attempting to extend the appeal of the package into the lucrative social science market.

11.5 LIMITATIONS

The advantages offered by generalised linear models should not persuade geographers that there are no problems or limitations in their use. Geographical problems are sometimes especially difficult to handle statistically and require tailor-made techniques. As generalised linear models were not designed with geography in mind it is not surprising that they are of limited value for certain types of geographical data handling.

There are three main areas of limitation which should be recognised:

1 they are not a substitute for informed thought;
2 they are not a substitute for exploratory data analysis;
3 they are not a solution for all types of statistical problem.

11.5.1 Informed thought

It is tempting for geographers faced with a large and complex data set to let the computer package do the processing for them unaided. This temptation may be somewhat greater for users of generalised linear models because their design, and ready computation using GLIM, makes it possible to specify any number of different types of analysis with relative ease. There is thus a degree of risk involved in encouraging students to experiment with alternative types of modelling because they may use the package solely as a black box, i.e., producing a range of models which they cannot interpret.

The black box strategy can be valuable in situations where the potential number of alternative models is excessively large and a systematic step-by-step specification would be tedious and time-consuming. In this approach the computer may be set to fit a wide range of different types of model, including or excluding specific variables, transforming some, altering the functional form of the relationship etc., and merely reports a range of fitness values at the end. These could then be used to root out alternatives which are undesirable on the basis of the calculated measure, allowing

319

more time to examine in depth the subset of desirable and semi-desirable models.

The main drawback with this idea, which is popular in some geographical information system circles, is that the summary measure may be of little value for describing the patterns and relationships in the data. This is simply a restatement of the problem associated with correlation in which identical summary values may be generated by wholly different types of data. In other words, the subset chosen for further analysis may contain a widely diverse selection of models even though they appear to share a common summary measure for goodness-of-fit. Given that time will have to be spent investigating these in depth, it is arguable that the strategy is more comprehensive or time saving than the development of a final model from the preliminary analysis of the data.

What generalised linear models provide is a framework for relating different variables in an algebraic model. Ideally, the composition of this model should reflect what is already known from other studies about the relationships between the variables. This is because some attempt should be made to check that the data either conform to existing knowledge or differ in some specific ways. A research strategy based on such prior information can avoid the black box approach entirely as an informed reading of the existing literature may provide the necessary starting points for the modelling.

11.5.2 Exploratory data analysis

A second limitation of data handling using generalised linear models is that they do not accommodate difficult data without intervention by the geographer. This problem is common to all computer systems, not just GLIM, as they are rarely programmed to provide checks on the validity of the data.

All statistical models based on probability processes are designed to operate under specified conditions. These conditions are the assumptions underlying the model. For example, the behaviour of the regression model depends on the general validity of several different assumptions affecting the specification of the model, the magnitude of the data, and the characteristics of the random error term. These assumptions are made so that acceptable estimates of its parameters may be generated. For inferential purposes, it is also usual to assume that the error term has been drawn from a Normal distribution. If these assumptions are violated the model will behave suboptimally and the estimates will not be as good as they could be.

The only way of checking the general validity of these assumptions in any given data set is to carry out a preliminary analysis of the data beforehand to get a 'feel' for its structure. Then, having fitted the model, other summary measures and graphical devices (e.g., Goldfeld–Quandt test, partial residual plots, etc.) may be generated to investigate specific departures from these assumptions. Generalised linear models are as prone to assumption violations as the original models and so therefore need to be checked.

11.5.3 Statistical limitations

The third set of limitations concerns the fact that generalised linear models are not wholly appropriate for certain types of modelling problem. Three main areas of difficulty exist:

1 situations where models include non-linear parameters;
2 situations where there is more than one error component;
3 situations where mean–variance relationships are not constant or the observations are correlated.

This does not mean that generalised linear models cannot be devised for these types of problem. Usually, they can, but the analysis is generally more complex.

To illustrate this, consider the concept of linearity which is central to the development of this class of models. The term 'linearity' refers to the organisation of the linear predictor, i.e., to the structural form of the independent variables, co-variates and factors in the model. It is based on a generalisation of the idea used in regression in which the model is said to be 'linear-in-parameters'. Non-linearity therefore occurs when parameters are not linearly related to the linear predictor.

The introduction of non-linear parameters complicates the fitting algorithm used to obtain parameter estimates for generalised linear models and may, if their effect is sufficiently severe, cause it to fail to converge. These difficulties may, however, be overcome if the non-linear parameters can be transformed or linearised in some way. Alternatively, if the value of the non-linear parameter is known before analysis or can be fixed to a given level, the model may still be fitted as though it were linear. McCullagh and Nelder (1983, Chapter 10) gives further details.

11.6 THE FUTURE

Generalised linear models are now well established within the statistics and social science literature. Many have come to recognise the value of their integrated and flexible framework, as well as the possibilities they offer for powerful data analyses. Geography, with its feet set firmly in the econometrics tradition, has yet to make the breakthrough. It is to be hoped that by using generalised linear modelling many geographers will gain a wider and more catholic understanding of the problems of data analysis and modelling. Such an appreciation is absolutely vital for geography because of the range of data types used in our work and the vast quantities of data which are available. The need for a solid grounding in a wider model of statistics teaching than is usually practised becomes all the more important as geographers develop geographical information systems. These bring together extensive data collections with map-making facilities and query languages. Without an understanding of where the data come from and how they relate both statistically and spatially the great potentials of these systems will not be realised. Spatial analysis requires flexible statistical analysis. Generalised linear models provide that flexibility.

APPENDIX A

THE EXPONENTIAL FAMILY OF PROBABILITY DISTRIBUTIONS

A.1 MEMBERS OF THE EXPONENTIAL FAMILY

The justification for the development and presentation of many linear models as generalised linear models depends on theoretical findings from distribution theory. Many of the most commonly-used theoretical frequency distributions which underlie these models rely on probability processes which are themselves inter-related. It can be shown that the probability processes presented in this book are all members of a common family: the exponential family of probability distributions. Findings relevant to this family provide the inferential integration needed to develop GLIM.

The exponential family has been studied for fifty years or more, being discussed in the writings of, among others, Fisher (1935), Darmois (1935), Koopman (1936), Pitman (1936) and Lehmann (1959). However, many of the practical advantages had to await the development and wider use of powerful computers, capable of providing robust numerical approximations. More recent treatments of the exponential family can be found in Barndorff-Nielsen (1978) and Andersen (1980).

A.2 CHARACTERISTICS OF THE EXPONENTIAL FAMILY

The exponential family of probability distributions is characterised by the following equation (after Dobson 1983):

$$f(y;\theta) = s(y)t(\theta)e^{a(y)\,b(\theta)} \tag{A.1}$$

where

a, b, s, t are known functions

In this form, the probability function associated with discrete distributions, and the probability density function associated with continuous distributions, depends on only one paramater, θ. This may be made clearer by rewriting equation A.1 as:

$$f(y;\theta) = \exp\left[a(y)b(\theta) + c(\theta) + d(y)\right] \tag{A.2}$$

where

$s(y) = \exp d(y)$
$t(\theta) = \exp c(\theta)$
$a(y)$ is assumed to equal y, a random observation on response variable Y

The component $b(\theta)$ reflects the typical, or canonical, parameter of the distribution which distinguishes it from other members of the family. The three other components are treated as known nuisance parameters.

It is relatively easy to re-express the traditional formulae for commonly-used probability distributions in exponential family form. For example, the Poisson formula is

$$f(y; \lambda) = \frac{\lambda^y e^{-\lambda}}{y!} \text{ for } y = 0, \ldots, n \tag{A.3}$$

which becomes

$$f(y; \lambda) = \exp[y \log \lambda - \lambda - \log y!] \tag{A.4}$$

on modification. The canonical parameter for this distribution in $\log \lambda$. Similarly, the binomial and the Normal distributions:

$$f(y; \pi) = \binom{n}{y} p^\pi (1 - p)^{n-y} \qquad y = 0. \ldots, n \tag{A.5}$$

$$f(y; \mu) = \frac{1}{\sqrt{(2\pi\sigma^2)}} \exp\left[-\frac{1}{2\sigma^2} (y - \mu)^2 \right] \tag{A.6}$$

become

$$f(y; \pi) = \exp[y \log \pi - y \log(1 - \pi) + n\log(1 - \pi) + \log \binom{n}{y}] \tag{A.7}$$

$$f(y; \mu) -\exp\left[-\frac{y^2}{2\sigma^2} + \frac{y\mu}{\sigma^2} - \frac{\mu^2}{2\sigma^2} - \tfrac{1}{2} \log(2\pi\sigma^2) \right] \tag{A.8}$$

on modification. The canonical parameters of these distributions are

$$n \log(1 - \pi) \tag{A.9}$$

and

$$-\tfrac{1}{2} \mu^2/\sigma^2 - \tfrac{1}{2} \log(2\pi\sigma^2)$$

respectively.

A.3 MAXIMUM LIKELIHOOD ESTIMATION

The procedure which is applied to the data to generate maximum likelihood estimates for their parameters varies if there is insufficient independent information in the raw data. Two situations may be identified depending on the nature of the data matrices:

1 Analyses based on full rank matrices – a situation in which sufficient information exists.
2 Analyses based on matrices which are less than full rank – a situation in which insufficient independent information exists.

For analyses based on the former, the following relationships may be specified to permit estimation. First, the likelihood function with respect to β is obtained. Second, maximising for β produces the following equation:

$$A\beta = r \tag{A.10}$$

whose solution is given by:

$$\hat{\beta} = A^{-1} r \tag{A.11}$$

where

$$A = (X^T V^{-1} X),$$
$$r = X^T V^{-1} Z$$

The components of these equations are respectively:

X a full rank design matrix
V^{-1} a diagonal matrix of iterative weights
Z_i a random variable whose expectation is η and whose variance is V

The algebraic form of Z_i indicates how the algorithm improves the accuracy of the estimation:

$$Z_i = \eta_i + (y_i - \mu_i).\frac{\delta_\eta}{\delta_\mu} \tag{A.12}$$

This is equivalent to the minimisation of a weighted difference function.

From existing theory it can be shown that the estimates produced at convergence are maximum likelihood estimates. These are justified asymptotically, which implies that

$$\begin{aligned}
\text{var } (\beta) &= (X^T V X)^{-1} \\
\eta &= X\beta \\
\mu &= \text{inverse } (\hat{\eta})
\end{aligned} \tag{A.13}$$

A complication arises if X is of less than full rank because then there is no unique solution to equation A.3. Instead of A.4 the generalised solution:

$$\beta^* = A^{-*} r \tag{A.14}$$

is written. In this A^{-*} is any generalised inverse matrix of A. This means that several different types of matrix will provide possible solutions to equation A.14 (Holt 1979). In order to solve this equation therefore, some constraints need to be applied to reduce the solution space to a single point. Searle (1971) notes that a variety of systems of constraints exist which yield values of β which would be acceptable, and that in practice it is not clear which is to be preferred. Baker and Nelder (1978, Part 1:2) note:

> for statistical purposes the particular ... (system) ... chosen is irrelevant since, for example, $\hat{\eta}$, and thus $\hat{\mu}$, can be shown to be unique whichever value of β^* we choose. [The] choice of a particular β^* does not change the model we are fitting but merely determines our manner of expressing the linear structure.

This raises an important issue: for non-full-rank models it is inadvisable to interpret the individual parameters from the model as they reflect the constraints used. Consequently, they will vary in value even though the estimates of $\hat{\eta}$ and $\hat{\mu}$ will be unaffected. Holt (1979) provides a convincing illustration of the issue using a common data set but different systems of constraints. The issue is also considered at length in Wrigley (1985).

A data set which is of less than full rank can arise for two distinct reasons. On the one hand, it may occur because the sample size is too small to estimate all the parameters which could be applied. On the other, it could arise because the specification of the linear structure of the population is redundant whatever sample is selected. The former problem is termed extrinsic aliasing, the latter intrinsic aliasing. An illustration of the former is provided by a design matrix consisting of three variables, X_1, X_2, X_3, in which X_3 is a linear combination of the other two. Once X_1 and X_2 are known, it follows that X_3 is also known. Of the three parameters which could be used to represent these, only two are actually needed. This suggests how a possible system of constraints could be developed:

1 Use two parameters, setting the third to zero (a form of corner-weighting).
2 Use some form of summation criterion which ensures that the sum of the three parameters is zero (a form of centre-weighting).

GLIM employs the former system but sets the first parameter in each parameter effect to zero. All parameters are interpreted, if required, with respect to this anchor or bench-mark parameter.

The latter system, one version of which is called the usual constraints (Payne 1977), is used in many other programs, and is most readily met in the published literature, as it is based on the procedures used in the analysis of variance. Using this, parameters are interpreted with reference to the overall average of the expected cell frequencies, rather than an anchor cell. Consequently, it is no surprise to find that parameter estimates will differ depending on which system has been used. However, estimable functions based on them, for example odds-ratios, cross-product ratios, fitted values, will remain unaffected. This is because any solution $\hat{\beta}$ to the estimating equations will correspond to the same unique maximum.

APPENDIX B

OTHER SOFTWARE FOR MODELLING AND GEOGRAPHICAL DATA ANALYSIS

Throughout this book considerable attention has been given to the GLIM computer system for generalised linear models. Of all the many computer packages on the market, GLIM is perhaps the most suitable for a course in statistical analysis designed around the Nelder and Wedderburn approach. The principal relationships of this family of models lie at the core of GLIM and the command language has been designed to reinforce them.

However, there are many other packages available which can be used instead of GLIM. Some of these emphasise an alternative approach to modelling (for example, the GENCAT package is based on a framework of generalised chi-square analysis), while others merely provide a selection of independent and semi-independent modules associated with a particular technique or family of techniques.

Comparative reviews of some of these alternatives are provided in O'Brien and Wrigley (1980), Francis (1981) and Wetherill and Curram (1984, 1985). O'Brien (1986) provides some general information on the statistical software available for use with microcomputers.

B.1 MINITAB

A powerful computing package which is most suited to teaching uses and small data sets. Facilities exist for preliminary data analysis and data manipulation. Regression, analysis of variance and time-series analysis facilities are available. Categorical data facilities are limited.

B.2 BMDP

A general purpose statistics package designed initially for biomedical applications. Principally a batch-mode program, but can be used interactively. It accommodates user-supplied FORTRAN routines. Its facilities for data management are relatively primitive.

B.3 GENSTAT

A flexible general purpose package which has many features of a high-level programming language. Comprehensive range of options to suit the series analyst, but is hampered by a terse manual.

B.4 SAS

An extremely powerful and comprehensive system of modules and procedures which can accommodate virtually any type of data. Performs a wide variety of types of analysis. Is principally of use for data management, statistical analysis and report writing.

B.5 SPSSx

The successor to SPSS. This is a powerful series of procedures designed to facilitate 'production' analyses of survey data, rather than model fitting or exploration. SPSSx is widely available, and can be used with microcomputers.

B.6 GENCAT

A package designed to fit a comprehensive system of linear models based on the work of Grizzle *et al.* (1969). The package can handle all of the major linear statistical models in current geographical use but requires the user to supply the design matrices. This can be complex, see Wrigley (1980).

APPENDIX C

A SUMMARY OF THE GLIM 3 COMMAND LANGUAGE

C.1 INTRODUCTION

The following is a brief description of the command language used in GLIM 3, that is, the third release of the GLIM system. GLIM 3 was released in 1978 (see Baker and Nelder 1978 for details) and has since been significantly enhanced. Over one thousand copies of GLIM 3.12 have been installed world-wide on university and commercial computer systems. In 1986, NAG released GLIM 3.77, a substantially modified version of the system written in FORTRAN 77. In this, the range of facilities available has been considerably extended, particularly by the addition of new commands to allow GLIM data structures to be accessed from external FORTRAN command files, to allow preparatory and descriptive statistical analysis to be performed prior to the fitting of models, and to simplify many features of the otherwise terse command language. In addition, control over generated output has been improved with the addition of various facilities to adjust its style. Full details of GLIM 3.77 may be obtained in a significantly enlarged and structured manual (Payne 1986). Details of the command language are given in Baker *et al.* (1986: 151–65) and Baker (1986: 39–80). Details of incompatibilities between GLIM 3.12 and GLIM 3.77 are given in Baker (1986: 3–4).

C.2 COMMANDS

The following is a list of the commands available for use in GLIM 3 based on the list supplied in Baker and Nelder (1978) and Baker *et al.* (1986). Items starred with an asterisk are additions to the GLIM 3.12 commands which are only available for use in GLIM 3.77. Most of these new commands are designed to extend the range of graphical and data exploration facilities within the system.

COMMAND	DESCRIPTION
ACCURACY	Sets the number of significant figures to be used with printed output (default is 4).
ALIAS	Alters which parameters are to be intrinsically aliased from the default configuration (this influences the interpretation of the parameter estimates).
ARGUMENT	Assigns arguments to a macro.
*ASSIGN	Assigns a list of values to a vector.
CALCULATE	Allows the calculation of arithmetic expressions using, for example, simple operators, functions, and monodic functions.

COMMENT	Allows the insertion of comments into the program to aid interpretation. Comment lines are ignored in processing.
CYCLE	Controls the performance of the GLIM algorithm for model fitting.
DATA	Defines the labels to be used to describe variables in the current GLIM run.
DELETE	Deletes the values of specified variables. If the variable name has appeared in a DATA command, DELETE causes it to become undefined.
DINPUT	Allows the reading of data values from an external file.
DISPLAY	Displays a variety of types of output from a fitted model, e.g., parameter estimates and their standard errors. Output is directed by twelve options which are selected by letter.
DUMMY	This is actually a symbol (usually $) which is used to indicate the end of a previous statement.
DUMP	Causes the current state of the run to be written to an external file for future reference or re-use.
ECHO	A dummy switch which reverses its current state. It is used to print lines typed at a VDU or read in from an external file allowing you to check for inaccuracies. If 'on', an ECHO command sets printing 'off' (and vice versa).
EDIT	Allows the altering of one or more values within a vector.
END	Identifies the end of the current job. Control remains in GLIM for further analysis of a new job or a return to the operating system using STOP.
ENDMAC	Signifies the end of a macro. Control returns to the main GLIM program for further processing.
ENVIRONMENT	Provides information on the current state of the program. Output is governed by a set of options.
ERROR	Sets the error (probability) distribution to be used in the next model fit. Default options are available, as is a facility to generate your own, subject to certain restrictions.
EXIT	Allows premature termination of a macro (i.e., skipping over commands to the end).
EXTRACT	Copies the variance–covariance matrix and parameter estimates from the working matrix to system vectors.
FACTOR	Defines categorical variables (termed FACTORS) and their associated levels to be used within GLIM.
FINISH	An end-of-file marker used on secondary subfiles.
FIT	Causes GLIM to fit a statistical model to data using existing definitions of ERROR, YVARIABLE, LINK (or OWN), WEIGHT, OFFSET and SCALE.
FORMAT	Allows data items to be read into GLIM using FORTRAN formatting (fixed and free format available).
*GRAPH	Is used to produce graphical output. The exact function of this command depends on the computer installation.
*GROUP	Is used to categorise a continuous variable into discrete groupings, for plotting, description or analysis.
HELP	A dummy switch which reverses the current state of help messages after faults have been encountered. The quality of information returned (including suggestions for possible remedies) depends on existing state.

330

*HISTOGRAM	Allows the production of a histogram of specified variables. The format of the plot may be altered by options.
INPUT	Allows data and command lines to be read from an external file.
*LAYOUT	Like GRAPH, the function and format of this command depends on the computer installation. Usually used to control format of graphical plots.
LINK	Declares the link between η and μ in subsequent models. Eight standard options are available.
LOOK	Displays the current values of variables stored within GLIM.
LSEED	Invokes the local pseudo-random number generator (if attached).
MACRO	Defines a macro: a subset of commonly used, general purpose commands which need only be written once and are called as necessary.
*MANUAL	The exact function of this command depends on the installation. Its implied function is to print subsets of the current GLIM 3.77 manual.
*MAP	Is used to recode a range of data values on to specified values, e.g., midpoints, theoretically-defined minima and maxima.
OFFSET	Fixes the value of a variable to a prior value before analysis.
OUTPUT	Is used to define the output channel for printing or storage on a secondary attached file.
OWN	Allows the definition of a user-generated, i.e., not a default, generalised linear model.
*PAGE	A dummy switch to reverse the current state of pagination (i.e., 'on' or 'off').
*PASS	Allows access to GLIM data structures from user-supplied FORTRAN routines.
PAUSE	Produces a temporary return to the operating system, i.e., without destroying the current GLIM definitions and declarations.
PLOT	Is used to generate scatter plots. The form of these may be manipulated using five options.
PRINT	Generates printed output selected by an item list.
READ	Causes data for variables defined in a previous DATA command to be read into GLIM.
RECYCLE	(Like CYCLE) controls the performance of the GLIM algorithm for model fitting. Uses previous fitted values as new estimates in subsequent fits of standard models.
REINPUT	(As INPUT) except input begins from first record of file.
RESTORE	Restores existing GLIM program dumped to secondary file.
RETURN	Returns control from a secondary file back to GLIM.
REWIND	Rewinds an attached secondary file to its first record. Subsequent reads from this will begin at first record.
SCALE	Is used to calculate the scale parameter in modelling fitting. Defaults are set for certain models.
*SET	Is used to specify the mode of operation, i.e., batch mode or interactive.
SKIP	(Similar to EXIT) controls premature exiting from macros. It differs from EXIT in its mode of operation.

331

SORT	Permits ranking of vectors.
SSEED	Invokes the standard (i.e., non-local) pseudo-random number generator.
STOP	Terminates the current GLIM session and returns control to the operating system.
SUBFILE	A file containing distinct subsets of data and commands. Each subset is blocked by beginning and terminating commands which distinguish it from others.
SUSPEND	Returns control temporarily from a macro or secondary file to the main GLIM program.
SWITCH	A conditional switch which invokes one of a series of specified macros depending on the result.
*TABULATE	Produces tabulations of descriptive statistics or contingency tables.
*TPRINT	Produces tabular presentations of one or more vectors.
*TRANSCRIPT	Provides a log of the current run. Items logged may be varied by option list.
UNITS	Defines standard length for vectors to be used in subsequent analyses.
USE	Invokes the use of a macro.
VARIATE	Declares the name and length of variables explicitly.
*VERIFY	Dummy switch which reverses the current state of macro verifications. If 'on', each line read from a macro is written to current output channel before it is executed.
WARN	Dummy switch which reverses the current state of printing of warning messages.
WEIGHT	Defines prior weights for selected variables prior to fitting models.
WHILE	Allows repeated use of macro statements until a preset scalar attains zero.
YVARIABLE	Declares which variable is to be treated as the 'response' in subsequent models.

C.3 INCOMPATIBILITIES

The following incompatibilities exist between GLIM 3.12 and GLIM 3.77:

1 TERMS directive: this is no longer available in GLIM 3.77.
2 The DUMMY directive is no longer restricted to a single symbol. This is designed to aid the portability of the system.
3 The DUMP command has been reorganised. Files dumped from GLIM 3.12 may not be read by GLIM 3.77 unless reorganised into a suitable format.
4 The scale limits on the PLOT directive are computed differently in GLIM 3.77 using only the non-zero data points rather than all the data points.
5 The argument list associated with the USE command has been altered.

C.4 OTHER CHANGES

Three types of change have been made to the GLIM system with the introduction of GLIM 3.77. Most of these have been made to increase control over generated output and to simplify the more terse and inaccessible parts of the command language:

1 *Computational changes:* Changes have been made to the ACCURACY, CYCLE and RECYCLE commands to alter the tolerance limits for convergence, to aid the detection of intrinsic aliasing, and to alter the number of significant figures produced as output.

2 *Modifications to output:* Changes have been made to the DISPLAY, ENVIRON-MENT, FIT, LOOK, PLOT and PRINT commands to improve the quality and readability of output.

3 *Removal of restrictions:* Changes have been made to the INPUT, REINPUT, FORMAT, DATA and CALCULATE commands to remove unnecessary restrictions and to simplify them. CALCULATE in particular has been simplified by the development of a simple form of specification, the addition of new relational and logical operators, and the updating of the operator precedences.

BIBLIOGRAPHY

Adams, E. (1966) 'On the nature and purpose of measurement', *Synthese* 16, 152–69.

Agar, M.H. (1986) *Speaking of Ethnography: Qualitative Research Methods 2*, London, SAGE.

Agnew, J.A. and Cox, K.R. (1980) 'Immigration field dynamics: a conceptualism and field analysis', *Geografisker Annaler* 62B, 69–80.

Agresti, A. (1980) 'Generalised odds ratios for ordinal data', *Biometrics* 36, 59–67.

Agresti, A. (1984) *Analysis of Ordinal Categorical Data*, New York, Wiley.

Aitkin, M.A. (1978) 'The analysis of unbalanced cross-classifications (with discussion)', *Journal of the Royal Statistical Society*, A 141, 195–223.

Aitkin, M.A. (1979) 'A simultaneous test procedure for contingency table models', *Applied Statistics* 28, 233–42.

Aitkin, M.A. (1980) 'A note on the selection of log-linear models', *Biometrics* 36, 173–8.

Aitkin, M.A., Anderson, D., Francis, B. and Hinde, J. (1989) *Statistical Modelling in GLIM*, Oxford, Oxford University Press.

Alden, J. (1977) 'The extent and nature of double job holding in Great Britain', *Industrial Relations Journal* 8(3), 14–31.

Andersen, E.B. (1980) *Discrete Statistical Models with Social Science Applications*, Amsterdam, North Holland.

Anderson, E.W. (1977) 'Soil creep: an assessment of certain controlling factors with special reference to Upper Weardale, England', unpublished PhD thesis, Department of Geography, University of Durham.

Anderson, E.W. (1988) 'Dew measurements: eastern prosopis sp belt in the Wahiba Sand Sea', *Journal of Oman Studies* Special Report 3.

Anderson, E.W. and Cox, N.J. (1986) 'An assessment of soil movement time series from Brandon, County Durham, UK', *Z. Geomorph NF SupBD* 58, 145–54.

Anderson, E.W. and O'Brien, L.G. (1988) 'Space-time variability in dewfall: some findings from Oman', paper available from first author, Department of Geography, University of Durham.

Anderson, R.L. and Bancroft, T.A. (1952) *Statistical Theory in Research*, New York, McGraw-Hill.

Arrow, K.H. (1951) *Social Choice and Individual Value*, New York, Wiley.

Atkinson, A.C. (1981) 'Two graphical displays for outlying and influential observations in regression', *Biometrika* 68, 13–20.

Avadhani, M.S. and Sukhatme, B.V. (1965) 'Controlled simple random sampling', *Journal of the Indian Society of Agricultural Statistics* 17, 34–42.

Avadhani, M.S. and Sukhatme, B.V. (1968) 'Simplified procedures for designing controlled simple random sampling', *Australian Journal of Statistics* 10, 1–7.

Avadhani, M.S. and Sukhatme, B.V. (1973) 'Controlled sampling with equal probabilities and without replacement', *International Statistical Review* 41, 175–82.

Baird, J.C. (1970) *Psychophysical Analysis of Visual Space*, Oxford, Pergamon.

Baker, R.J. (1986) 'Reference Guide', Part 3 of C.D. Payne (ed.) *The GLIM System: Release 3.77*, Oxford, Numerical Algorithms Group Ltd.

Baker, R.J., Clarke, M.R.B., Francis, B., Green, M., Nelder, J.A., Payne, C.D., Reese, P.A. and Webb, J. (1986) 'Users Guide', Part 2 of C.D Payne (ed.) *The GLIM System: Release 3.77*, Oxford, Numerical Algorithms Group Ltd.

Baker, R.J. and Nelder, J.A. (1978) *The GLIM System: Release 3*, Oxford, Numerical Algorithms Group Ltd.

Barndorff-Nielsen, O. (1978) *Information and Exponential Families in Statistical Theory*, New York, Wiley.

Barndorff-Nielsen, O. and Cox, D.R. (1984) 'The effect of sampling rules on likelihood statistics', *International Statistical Review* 52, 309–26.

Bartholomew, D.J. (1981) *Mathematical Models in Social Science*, Chichester, Wiley.

Bartlett, M.S. (1935) 'Contingency table interactions', *Journal of the Royal Statistical Society* Supplement 2, 248–52.

Bates, J.J. (1986) 'Stated preference techniques and the analysis of consumer choice', in N. Wrigley (ed.) *Store Choice, Store Location and Market Analysis*, London, Routledge.

Bateson, N. (1984) *Data Construction in Social Surveys*, London, Allen & Unwin.

Batty, M. (1976) *Urban Modelling: Algorithms, Calibration, Prediction*, Cambridge, Cambridge University Press.

Batty, M., Bracken, I., Guy, C.M. and Spooner, R. (1985) 'Teaching spatial modelling using interacting computers and interactive computer graphics', *Journal of Geography in Higher Education* 9, 25–36.

Baxter, M.J. (1979) 'The application of logit regression analysis to production-constrained gravity models', *Journal of Regional Science* 19, 171–7.

Baxter, M.J. (1983) 'Estimation and inference in spatial interaction models', *Progress in Human Geography* 7, 40–59.

Belsley, D.A., Kuh, E. and Welsh, R.E. (1980) *Regression Diagnostics: Identifying Influential Data and Sources of Collinearity*, New York, Wiley.

Ben-Akiva, M. and Lerman, S. (1985) *Discrete Choice Analysis: Theory and Applications to Travel Demand*, Cambridge, Mass., MIT Press.

Benedetti, J.K. and Brown, M.B. (1978) 'Strategies for the selection of log-linear models', *Biometrics* 34, 680–6.

Besag, J. (1986) 'On the statistical analysis of dirty pictures', *Journal of the Royal Statistical Society, B* 48, 259–302.

Bhapkar, V.P. (1966) 'A note on the equivalence of two test criteria for hypotheses in categorical data', *Journal of the American Statistical Association* 61, 228–35.

Bibby, J. (1977) 'The general log-linear model – a cautionary tale', in C.A. O'Muircheartaigh and C.D. Payne (eds) *The Analysis of Survey Data*, vol. 2, London, Wiley.

Birch, M.W. (1963) 'Maximum likelihood in three-way contingency tables', *Journal of the Royal Statistical Society B* 25, 220–33.

Bishop, Y.M.M., Fienberg, S.E. and Holland, P.W. (1975) *Discrete Multivariate Analysis: Theory and Practice*, Cambridge, Mass., MIT Press.

Blalock, H.M. (1979) *Social Statistics*, Tokyo, McGraw-Hill Kogakusha.

BIBLIOGRAPHY

Blaxter, M. (1976) *The Meaning of Disability*, London, Heinemann.

Blundell, R. (1988) 'Consumer behaviour: theory and empirical evidence – a survey', *Economic Journal* 98, 16–65.

Boden, P., Stillwell, J. and Rees, P. (1987) 'Migration data from the NHSCR and 1981 Census: further comparative analysis of aggregate inter-zonal information', Paper presented to the Population Geography Study Group Conference Information Systems for Populations and their Demographic, Socioeconomic and Housing Characteristics, Mansfield College, Oxford.

Bowlby, S. and Silk, J. (1982) 'Analysis of qualitative data using GLIM: two examples based on shopping survey data', *Professional Geographer* 34, 80–90.

Breen, R. (1985) 'Log-multiplicative models for contingency tables using GLIM', *GLIM Newsletter* 10, 14–19.

Brodsky, H. and Shalom Hakkert, A. (1985) 'Accessibility and bystander response in an emergency', *Transactions of the Institute of British Geographers* 10, 303–16.

Brown, M.B. (1976) 'Screening effects in multidimensional contingency tables', *Applied Statistics* 25, 37–46.

Brown, M.B. (1981) 'Module P4F', in W.J. Dixon (ed.) *BMDP: Biomedical Computer Programs*, University of California.

Brown, P.J. and Zidek, J.V. (1980) 'Adaptive multivariate ridge regression', *Annals of Statistics* 8, 64–74.

Burn, C. and Fox, M. (1986) 'Introducing statistics to geography students: the case for exploratory data analysis', *Journal of Geography* 28–31.

Burrough, P. (1986) *Principles of Geographical Information Systems for Land Resources Assessment*, Oxford, Clarendon Press.

Buse, A. (1982) 'The likelihood ratio, Wald and Lagrange multiplier tests: an expository note', *American Statistician* 36, 153–7.

Campbell, N.R. (1928) *An Account of the Principles of Measurement and Calculation*, London.

Carruthers, A.W. and Waugh, T.C. (1988) *GIMMS Reference Manual 5*, Edinburgh, Scotland. GIMMS Ltd, 30 Keir Street.

Chatfield, C. (1980) *The Analysis of Time Series*, London, Chapman and Hall.

Chatfield, C. (1982) 'Teaching a course in applied statistics', *Applied Statistics* 31, 272–89.

Chatfield, C. (1983) *Statistics for Technology*, London, Chapman and Hall.

Chatfield, C. and Collins, A.J. (1980) *An Introduction to Multivariate Analysis*, London, Chapman and Hall.

Chatterjee, S. and Price, B. (1977) *Regression Analysis by Example*, New York, Wiley.

Chrisman, N.R. (1981) 'Methods of spatial analysis based on maps of categorical coverages', unpublished PhD thesis, Department of Geography, University of Bristol.

Cliff, A.D. (1975) *Elements of Spatial Structure: A Quantitative Approach*, Cambridge, Cambridge University Press.

Cliff, A.D. and Ord, J.K. (1973) *Spatial Autocorrelation*, London, Pion.

Cliff, A.D. and Ord, J.K. (1981) *Spatial Processes*, London, Pion.

Cochran, W.G. and Cox, G.M. (1957) *Experimental Designs*, 2nd edn, New York, Wiley.

Cohen, L. and Holliday, M. (1982) *Statistics for Social Scientists*, London, Harper & Row.

Coleman, A. (1961) 'The second land use survey: progress and prospect', *Geographical Journal* 127, 168–86.

Collett, D. (1979) 'Review of GLIM 3 Users Manual', *Biometrics* 35, 327–8.

Cook, R.D. and Weisberg, S. (1982) *Residuals and Influence in Regression*, London, Chapman and Hall.

Coombs, C. (1950) 'Psychological scaling without a unit of measurement', *Psychological Review* 57, 145–58.

Cormack, R.M. (1980) 'Model selection in capture–recapture experiments', *Biometrics Unit Mimeo Series*, BU712M.

Cormack, R.M. (1981) 'Loglinear models for capture–recapture experiments on open populations', in R.W. Hiorns and D. Cooke (eds) *The Mathematical Theory of the Dynamics of Biological Population*, London, Academic Press.

Cox, D.R. (1970) *Analysis of Binary Data*, London, Chapman and Hall.

Cox, D.R. (1972) 'Regression models and life tables', *Journal of the Royal Statistical Society, B* 34, 187–220.

Cox, D.R. (1975) 'Partial likelihood', *Biometrika* 62, 269–76.

Cox, D.R. (1984) 'Interaction', *International Statistical Review* 52, 1–31.

Cox, D.R. and Snell, E.J. (1968) 'A general definition of residuals (with discussion)', *Journal of the Royal Statistical Society, B* 30, 248–75.

Cox, N.J. and Anderson, E.W. (1978) 'Teaching geographical data analysis: problems and possible solutions', *Journal of Geography in Higher Education* 2, 29–37.

Cox, N.J. and Jones, K. (1981) 'Exploratory data analysis', in N. Wrigley and R.J. Bennett (eds) *Quantitative Geography: A British View*, London, Routledge & Kegan Paul.

CURDS (1983) 'Functional regions: definitions, applications, advantages', Fact-sheet 1, Functional Regions Series, Centre for Urban and Regional Development Studies, University of Newcastle upon Tyne, Newcastle NE1 7RU.

Daganzo, C.F. (1979) *Multinomial Probit: The Theory and its Application to Demand Forecasting*, New York, Academic Press.

Daganzo, C.F., Bouthelier, F. and Sheffi, Y. (1977) 'Multinomial probit and qualitative choice: a computationally efficient algorithm', *Transportation Science* 11, 338–58.

Darmois, G. (1935) 'Sur les lois de probabilitie à estimation', *C.R. Acad. Sci*, Paris, 260, 1265–6.

Darroch, J.N. (1974) 'Multiplicative and additive interactions in contingency tables', *Biometrika* 61, 207–14.

Davie, R., Butler, N. and Goldstein, H. (1972) *From Birth to Seven: A Report of the National Child Development Study*', London, Longman.

Davies, R.B. (1987) 'Robustness in modelling dynamics of choice', in J. Hauer, H. Timmermans and N. Wrigley (eds) *Contemporary Developments in Quantitative Geography*, Dordrecht, D. Reidel.

Davies, R.B. and Guy, C.M. (1987) 'The statistical modelling of flow data when the Poisson assumption is violated', *Geographical Analysis* 19, 300–14.

Deaton, A.S. and Muellbauer, J. (1980) *Economics and Consumer Behaviour*, Cambridge, Cambridge University Press.

Defize, P.R. (1980) 'The calculation of adjusted residuals for log-linear models in GLIM', *GLIM Newsletter* 3, 41.

Denham, C. (1984) 'Urban Britain', *Population Trends* 36, 10–18.

Department of the Environment (1987) *Handling Geographic Information*, London, HMSO.

Devis, T. (1984) 'Population movements measured by the National Health Service Central Register', *Population Trends* 36, 18–24.

Devis, T. and Mills, J. (1986) 'A comparison of migration data from the National

Health Service Central Register and the 1981 Census', Occasional Paper 35, London, Office of Population Censuses and Surveys.

Dixon, C. and Leach, B. (1977) *Sampling Methods for Geographical Research, Concepts and Techniques in Modern Geography*, 17, Norwich, Geobooks.

Dixon, W.J. (ed.) (1983) *BMDP: Biomedical Computer Programs*, University of California Press.

Dobson, A. (1983) *An Introduction to Statistical Analysis*, London, Chapman and Hall.

Duncan, O. (1974) 'Footnotes', *Proceedings of the American Sociological Association*

Duncan, O.D. (1979) 'How destination depends on origin in the occupational mobility table', *American Journal of Sociology* 84, 793–803.

Dunn, R.J., Reader, S. and Wrigley, N. (1983) 'An investigation of the assumptions of the NBD model as applied to purchasing at individual stores', *Applied Statistics* 32.

Dunn, R.J. and Wrigley, N. (1984) 'Diagnostics and resistant fits in logit choice models', in D.E. Pitfield (ed.) *Discrete Choice Models in Regional Science*, London, Pion.

Dunn, R.J. and Wrigley, N. (1985) 'Beta-logistic models of urban shopping center choice', *Geographical Analysis* 17, 95–113.

Durbin, J. (1987) 'Statistics and statistical science', *Journal of the Royal Statistical Society, A* 150.

Edwards, A.W.F. (1972) *Likelihood*, Cambridge, Cambridge University Press.

Efroymson, M.A. (1960) 'Multiple regression analysis', in A. Ralston and H.S. Wilf (eds) *Mathematical Models for Digital Computers*, New York, Wiley.

Ehrenberg, A.S.C. (1972) *Repeat Buying: Theory and Applications*, Amsterdam.

Ehrenberg, A.S.C. (1975) *Data Reduction*, Chichester, Wiley.

Ehrenberg, A.S.C. (1982) *A Primer in Data Reduction*, Chichester, Wiley.

Ellis, B. (1966) *Basic Concepts of Measurement*, Cambridge, Cambridge University Press.

Fay, R.E. and Goodman, L.A. (1975) *The ECTA Program: Description for Users*, Department of Statistics, University of Chicago.

Fienberg, S.E. (1972) 'The multiple recapture census for closed populations and incomplete 2k contingency tables', *Biometrika* 59, 591–603.

Fienberg, S.E. (1977) *The Analysis of Cross-classified Categorical Data*, Cambridge, Mass., MIT Press.

Fienberg, S.E. and Mason, W. (1979) 'Identification and estimation of age, period and cohort models in the analysis of discrete archival data', *Sociological Methodology* 1–67.

Fingleton, B. (1981) 'Log-linear models, mostellerizing and forecasting', *Area* 13, 123–9.

Fingleton, B. (1983) 'Independence, stationarity, categorical spatial data and the chi-square test', *Environment and Planning A*, 15, 483–99.

Fingleton, B. (1984) *Models of Category Counts*, Cambridge, Cambridge University Press.

Fisher, R.A. (1935) 'The case of zero survivors', *Annals of Applied Biology* 22, 164–5.

Fleiss, J.L. (1981) *Statistical Methods for Rates and Proportions*, 2nd edn, New York, Wiley.

Flohn, N. (1950) 'Neue Anschauungen uber die allgemeine Zirkulation der Atmosphäre und ihre klimatische Bedeutung', *Erdkunde* 11, 161–75.

Flowerdew, R. and Aitkin, M.A. (1982) 'A method of fitting the gravity model based on the Poisson distribution', *Journal of Regional Science* 22, 191–202.

Foot, D. (1981) *Operational Urban Models: An Introduction*, Andover, Methuen.

Fotheringham, A.S. (1983) 'A new set of spatial interaction models: the theory of competing destinations', *Environment and Planning, A* 15, 15–36.

Fotheringham, A.S. (1984) 'Spatial flows and spatial patterns', *Environment and Planning, A* 16, 529–43.

Fotheringham, A.S. (1985) 'Spatial competition and agglomeration in urban modelling', *Environment and Planning, A* 17, 213–30.

Fotheringham, A.S. and Webber, M.J. (1980) 'Spatial structure and distance-decay parameters', *Geographical Analysis* 12, 33–46.

Francis, I. (1981) *Statistical Software: A Comparative Review*, New York, North Holland.

Freeman, D.W. (1987) *Applied Categorical Data Analysis*, New York, Marcel Dekker.

Friedson, E. (1965) 'Disability as social deviance', in M.B. Sussman (ed.) *Sociology of Disability and Rehabilitation*, Washington DC, American Sociological Association.

Furnival, G.M. (1971) 'All possible regressions with less computation', *Technometrics* 13, 404–8.

Furnival, G.M. and Wilson, R.W.M. (1974) 'Regression by leaps and bounds', *Technometrics* 16, 499–511.

Garside, M.J. (1965) 'The best subset in multiple regression analysis', *Applied Statistics* 14, 196–200.

Gaudry, M.J.I. (1980) 'Dogit and logit models of travel mode choice in Montreal', *Canadian Journal of Economics* 13, 268–79.

Gaudry, M.J.I. and Degenais, M.G. (1979) 'The dogit model', *Transportation Research, B* 13, 105–11.

Gherardini, P.G. (1980) 'Interactive ridge regression with GLIM', *GLIM Newsletter* 2, 16–26.

Gilchrist, R. (1981) 'Calculation of residuals for all GLIM models', *GLIM Newsletter* 4, 26–8.

Gilchrist, R. (1983) 'GLIM syntax for adjusted residuals', *GLIM Newsletter* 6, 64–5.

Godambe, V.P. and Heyde, C.C. (1987) 'Quasi-likelihood and optimal estimation', *International Statistical Review*, 55, 231–44.

Goldstein, H. (1979) 'Specifying a multivariate logit model using GLIM', *GLIM Newsletter* 1, 23–6.

Goodman, L.A. (1968) 'The analysis of cross-classified data: independence, quasi-independence and interactions in contingency tables with or without missing entries', *Journal of the American Statistical Association* 63, 1091–131.

Goodman, L.A. (1970) 'The multivariate analysis of qualitative data: interactions among multiple classifications', *Journal of the American Statistical Association*, 65, 226–56.

Goodman, L.A. (1979a) 'Multiplicative models for the analysis of occupational mobility tables and other kinds of cross-classification table, *American Journal of Sociology* 84, 804–19.

Goodman, L.A. (1979b) 'Simple models for the analysis of association in cross-classifications having ordered categories', *Journal of the American Statistical Association* 74, 337–52.

Goodman, L.A. (1981) 'Association models and the bivariate normal distribution in the analysis of cross-classifications having ordered categories', *Biometrika* 68, 347–55.

Goodman, L.A. and Kruskal, W. (1954) 'Measures of association for cross-classifications', *Journal of the American Statistical Association* 49, 732–64.

Goodman, L.A. and Kruskal, W. (1959) 'Measures of association for cross-classifications 2: Further discussion and references', *Journal of the American Statistical Association*, 54, 123–63.

Goodman, L.A. and Kruskal, W. (1963) 'Measures of association for cross-classifications 3: Approximate sampling theory', *Journal of the American Statistical Association* 58, 310–64.

Goodman, L.A. and Kruskal, W. (1972) 'Measures of association for cross-classifications 4: Simplification of asymptotic variance', *Journal of the American Statistical Association* 67, 415–21.

Gourieroux, C., Monfort, A. and Trognon, A. (1984) 'Pseudo-maximum likelihood methods: theory', *Econometrica* 52, 681–700.

Gregory, C. and Altman, J. (1989) *Observing the Economy*, London, Routledge.

Gregory, D. (1981) 'Definition of accessibility', in R.J. Johnston (ed.) *The Dictionary of Human Geography*, Oxford, Blackwell.

Grizzle, J.E., Starmer, C.F. and Koch, G.G. (1969) 'Analysis of categorical data by linear models', *Biometrics* 25, 489–504.

Guy, C.M. (1983a) *Household Income and Food Shopping Behaviour*, Information Brief 83/5, Unit for Retail Planning Information Ltd, 20 Queen Victoria Street, Reading, RG1 1TG.

Guy, C.M. (1983b) 'The assessment of access to local shopping opportunities: a comparison of accessibility measures', *Environment and Planning, B* 10, 219–38.

Guy, C.M. (1984) *Superstore Shopping in Cardiff*, Information Brief 84/5, Unit for Retail Planning Information Ltd, 20 Queen Victoria Street, Reading, RG1 1TG.

Guy, C.M. (1987) 'Recent advances in spatial interaction modelling: an application to the forecasting of shopping travel', *Environment and Planning, A* 19, 173–86.

Guy, C.M. (1988) 'A unified statistical and computing environment for the analysis of shopping behaviour', Technical Reports in Geo-Information Systems, Computing and Cartography, Wales and South-West Regional Research Laboratory, 17, London.

Guy, C.M. (forthcoming) 'Spatial interaction modelling in retail planning practice: the need for robust statistical methods', *Environment and Planning, B* 18,.

Guy, C.M. and Broom, D. (1983) 'A disaggregate analysis of urban retail change', mimeo copy available from C.M. Guy, Department of Town Planning, University of Wales Institute of Science and Technology, Cardiff, Wales.

Guy, C.M. and O'Brien, L.G. (1983) 'Measurement of grocery prices: some methodological considerations and empirical results', *Journal of Consumer Studies and Home Economics* 7, 213–27.

Guy, C.M., Wrigley, N., O'Brien, L.G. and Hiscocks, G.K. (1983) 'The Cardiff Consumer Survey: a report on the methodology', Papers in Planning Research, 68, Department of Town Planning, UWIST, Cardiff, Wales.

Haberman, S.J. (1970) 'The general log-linear model', unpublished PhD thesis, Department of Statistics, University of Chicago.

Haberman, S.J. (1974a) *The Analysis of Frequency Data*, Chicago, University of Chicago Press.

Haberman, S.J. (1974b) 'Log-linear models for frequency tables with ordered classifications', *Biometrics* 30, 589–600.

Haberman, S.J. (1979) *Analysis of Qualitative Data*, vol. 2, *New Developments*, New York, Academic Press.

Hacking, I. (1965) *Logic of Statistical Inference*, Cambridge, Cambridge University Press.

341

Hacking, I. (1975) *The Emergence of Probability: A Philosophical Study of Early Ideas about Probability, Induction and Statistical Inference*, Cambridge, Cambridge University Press.

Haggett, P. (1975) *Geography: A Modern Synthesis*, New York, Harper & Row.

Haggett, P., Cliff, A.D. and Frey, A.E. (1977) *Locational Analysis in Human Geography*, London, Edward Arnold.

Haining, R. (1987) 'Geography and spatial statistics: current position, future developments', Paper presented at the Oxford Conference on New Models in Geography.

Hakim, C. (1982) *Social Analysis and Social Research*, London, Allen & Unwin.

Hamnett, C. and Randolph, W. (1987) 'The OPCS longitudinal study: a new tool for social research in England and Wales', *Area* 19, 69–73.

Hanuschek, E.A. and Jackson, J.E. (1977) *Statistical Methods for Social Scientists*, New York, Academic Press.

Harre, R. (1979) *Social Being*, Oxford, Blackwell.

Harrell-Bond, B. (1986) *Imposing Aid: Emergency Assistance to Refugees*, Oxford, Oxford University Press.

Harris, F.W. and O'Brien, L.G. (1988) 'The changing face of the town', *Geographical Magazine* 60, 34–7.

Harvey, D.W. (1973) *Social Justice and the City*, London, Edward Arnold.

Hausman, J.A. and Wise, D.A. (1978) 'A conditional probit model for qualitative choice: discrete decisions recognising interdependence and heterogeneous preferences', *Econometrica* 46, 403–26.

Haworth, J.M. and Vincent, P. (1979) 'The stochastic disturbance specification and its implications for log-linear regression', *Environment and Planning, A* 11, 781–90.

Haynes, K.M. and Fotheringham, A.S. (1984) *Gravity and Spatial Interaction Models*, Beverly Hills, Sage.

Healy, M.J.R. (1989) *GLIM: An Introduction*, Oxford, Oxford University Press.

Hensher, D.A. and Johnson, P.W. (1981) *Applied Discrete Choice Modelling*, London, Croom Helm.

Hewlett, P.S. and Plackett, R.L. (1978) *An Introduction to the Interpretation of Quantal Responses in Biology*, London, Edward Arnold.

Hoaglin, D.C. and Welsh, R.E. (1978) 'The hat matrix in regression and ANOVA', *American Statistician* 32, 17–22.

Hoerl, A.E. and Kennard, R.W. (1970) 'Ridge regression: biased estimation for non-orthogonal problems', *Technometrics* 12, 55–67.

Hoinville, G. and Jowell, R. (eds) (1978) *Survey Research Practice*, London, Heinemann.

Holman, J. (1980) 'Apprenticeship as a factor in migration: Bristol 1675–1716', *Bristol and Gloucester Architectural Journal* 97, 85–92.

Holt, D. (1979) 'Log-linear models for contingency table analysis', *Sociological Methods and Research* 7, 330–6.

Horton, R.L. (1978) *The General Linear Model*, New York, McGraw-Hill International.

Huff, D.L. (1963) 'A probabilistic analysis of shopping centre trade areas', *Land Economy* 39, 81–90.

Hutchinson, D.A. (1985) 'Ordinal variable regression using the McCullagh (proportional odds) model', *GLIM Newsletter* 9, 9–17.

Jacob, H. (1984) *Using Published Data: Errors and Remedies*, London, Sage.

Jensen, S. (1980) *Analysis of Intercity Travels by Railways in Denmark*, Research report 3, The Institute of Mathematics, Statistics and Operational Research, Technical University of Denmark, Lyngby, Denmark.

John, P.W.M. (1971) *Statistical Design and Analysis of Experiments*, New York, Macmillan.

Johnson, N.L. and Kotz, S. (1969) *Distributions in Statistics*, vol. 1, *Discrete Data*, New York, Wiley.

Johnson, N.L. and Kotz, S. (1970a) *Distributions in Statistics*, vol. 2, *Continuous Univariate Distributions 1*, New York, Wiley.

Johnson, N.L. and Kotz, S. (1970b) *Distributions in Statistics*, vol. 3, *Continuous Univariate Distributions 2*, New York, Wiley.

Johnson, N.L. and Kotz, S. (1972) *Distributions in Statistics* vol. 4, *Continuous Multivariate Distributions*, New York, Wiley.

Jones, F.L. and Pittelkow, Y.E. (1983) 'Analysis of occupational mobility tables using GLIM', *GLIM Newsletter* 7, 34–6.

Jones, K. (1981) 'Confirmatory and exploratory approaches to statistical inference in geography', Paper presented to the SSRC Quantitative Methods Conference, University of Bristol.

Jones, K. (1984) 'Graphical methods for exploring relationships', in G. Bahrenberg, M.M. Fischer and P. Nijkamp (eds) *Recent Developments in Spatial Data Analysis: Methodology, Measurement, Models*, Aldershot, Gower Publications.

Kalton, G., Rogers, J. and Holt, D. (1980) 'The effects of offering a middle option with opinion questions', *The Statistician* 29, 65–79.

Kennedy, P. (1979) *A Guide to Econometrics*, Oxford, Martin Robertson.

Keyfitz, N. (1945) 'The sampling approach to economic data', *Canadian Journal of Economics and Political Science* 11, 467–77.

Kiecolt, K. and Nathan, L. (1985) *Secondary Analysis of Survey Data*, Beverly Hills, Sage.

King, L.J. (1969) *Statistical Analysis in Geography*, Englewood Cliffs, Prentice Hall.

Kirk, J. and Miller, M. (1986) *Reliability and Validity in Qualitative Research: Qualitative Research Methods 1*, London, Sage.

Kish, L. (1967) *Survey Sampling*, New York, Wiley.

Kish, L. and Frankel, M.R. (1974) 'Inference from complex samples', *Journal of the Royal Statistical Society, B* 36, 1–37.

Koopman, L.H. (1936) 'On distributions admitting a sufficient statistic', *Transactions of the American Mathematical Society* 39, 399–409.

Kruskal, J.B. and Wish, M. (1978) *Multidimensional Scaling*, Beverly Hills, Sage.

Labovitz, S. (1967) 'Some observations on measurement and statistics', *Social Forces* 46, 151–60.

Labovitz, S. (1972) 'Statistical usage in sociology: sacred cows and ritual', *Sociological Methods and Research* 1, 13–37.

Lakshmanan, T.R. and Hansen, W.G. (1965) 'A retail market potential model', *Journal of the American Institute of Planners* 31, 95–108.

LAMSAC (1979) 'User specifications for a census data handling package', London, Local Authorities Management Services Advisory Committee, Census Working Party.

LAMSAC (1982) *SASPAC 2.5 User Manual*, London, Local Authorities Management Services Advisory Committee.

Lancaster, K. (1966) 'A new approach to consumer theory', *Journal of Political Economy* 74, 132–57.

Lancaster, K.J. (1971) *Consumer Demand: A New Approach*, New York, Columbia University Press.

343

BIBLIOGRAPHY

Lehmann, E.L. (1959) *Testing Statistical Hypotheses*, New York, Wiley.
Lewis, P. (1975) *Maps and Statistics*, London, Methuen.
Longley, P. and Wrigley, N. (1984) 'Scaling residential preferences: a methodological note', *Tijdschrift voor Economische en Sociale Geografie* 75, 292–9.
Lovett, A.A., Whyte, I.D. and Whyte, K.A. (1985) 'Poisson regression analysis and migration fields: the example of the apprenticeship records of Edinburgh in the 17th and 18th centuries', *Transactions of the Institute of British Geographers* 10, 317–32.
Luce, R.D. (1959) *Individual Choice Behaviour*, New York, Wiley.
Luce, R.D., Green, D.M. and Weber, D. (1976) 'Attention bands in absolute identification', *Perception and Psychophysics* 20, 49–54.

Manly, B.F.J. and Parr, M.J. (1968) 'A new method of estimating population size, survivorship, and birth rate from capture–recapture data', *Transactions of the British (TSBE)* 18, 81–9.
Manski, C.F. (1981) 'Structural models for discrete data: the analysis of discrete choice', in S. Leinhardt (ed.) *Sociological Methodology*, San Francisco, Jossey-Bass, pp. 58–109.
Manski, C.F. and McFadden, D. (eds) (1981) *Structural Analysis of Discrete Data with Econometric Applications*, Cambridge, Mass., MIT Press.
MAPICS (1986) *The MAPICS Reference Manual*, London, MAPICS Ltd.
Marsh, C. (1982) *The Survey Method*, London, Allen & Unwin.
Marsh, C. (1989) *Exploring Social Data*, Cambridge, Polity.
Massam, B. (1975) *Location and Space in Social Administration*, London, Edward Arnold.
Mather, P.M. (1976) *Computational Methods of Multivariate Analysis in Physical Geography*, Chichester, Wiley.
Mather, P. (1991) *Computer Applications in Geography*, London, Wiley.
McCullagh, P. and Nelder, J.A. (1983) *Generalised Linear Models*, London, Chapman and Hall.
McFadden, D. (1974) 'Conditional logit analysis of qualitative choice behaviour', in P. Zarembka (ed.) *Frontiers in Econometrics*, New York, Academic Press, pp. 105–42.
McFadden, D. (1976) 'Quantal choice analysis: a survey', *Annals of Economic and Social Measurement* 5, 363–90.
McKennall, A.C. (1977) 'Attitude scale construction', in C.A. O'Muircheartaigh and C. Payne (eds) *The Analysis of Survey Data*, vol. 1, London, Wiley.
Menges, G. (1973) 'Inference and decision', in *Selecta Statistica Canadiana*, vol. 1, Toronto, University Press Canada.
Meyer, R.J. and Eagle, T.C. (1981) 'A parsimonious multinomial choice model recognising alternative interdependence and context-dependent utility functions', Working Paper 26–80–81, Graduate School of Industrial Administration, Carnegie-Mellon University.
Miller, G.A. (1956) 'Information and memory', *Scientific American*,
Moser, C. and Kalton, G. (1971) *Survey Method in Social Investigation*, Aldershot, Gower.
Mosteller, F.W. (1968) 'Association and estimation in contingency tables', *Journal of the American Statistical Association* 63, 1–28.

Namboodiri, N.K., Carter, L.F. and Blalock, H.M. (1975) *Applied Multivariate Analysis and Experimental Designs*, New York, McGraw-Hill.
Nelder, J.A. (1974) 'Log-linear models for contingency tables: a generalisation of classical least-squares', *Applied Statistics* 23, 323–9.

344

Nelder, J.A. (1977) 'A reformulation of linear models', *Journal of the Royal Statistical Society, A* 140, 48–77.

Nelder, J.A. (1984) 'Statistical models for qualitative data', in P. Nijkamp, P. Leitner and N. Wrigley (eds) *Measuring the Unmeasurable: Analysis of Qualitative Spatial Data*, The Hague, Martinus Nijhoff, Chapter 2.

Nelder, J.A. and Wedderburn, R.W.M. (1972) 'Generalised linear models', *Journal of the Royal Statistical Society, A* 135, 370–84.

Neyman, J. (1949) 'Contributions to the theory of the chi-square test', in J. Neyman (ed.) *Proceedings of the 1st Berkeley Symposium on Mathematical Statistics and Probability*, Berkeley, University of California Press.

O'Brien, L.G. (1982) 'Categorical data analysis for geographical research: with applications to public sector residential mobility', unpublished PhD thesis, Department of Geography, University of Bristol.

O'Brien, L.G. (1983) 'Generalised linear modelling using the GLIM system', *Area* 15, 327–36.

O'Brien, L.G. (1986) 'Statistical software for microcomputers', *Area* 18, 39–42.

O'Brien, L.G. (1987a) 'User control versus randomisation in geographical probability sampling: a compromise solution using controlled sampling', *Environment and Planning, A* 19, 949–58.

O'Brien, L.G. (1987b) 'GLIM 3.77 software review', *Professional Geographer* 39, 29–30.

O'Brien, L.G. (1989) *The Statistical Analysis of Contingency Table Designs: Concepts and Techniques in Modern Geography*, 51, Norwich, Environmental Publications, University of East Anglia.

O'Brien, L.G. and Guy, C.M. (1985) 'Locational variability in retail grocery prices', *Environment and Planning, A* 17, 953–62.

O'Brien, L.G., Nelson, R., Dodds, P. and Blakemore, M.J. (1987) *NOMIS User Manual*, vols 1–4, *National Online Manpower Information System*, Department of Geography, University of Durham, UK.

O'Brien, L.G. and Wrigley, N. (1980) 'Computer software for the analysis of categorical data', *Area* 12, 263–8.

O'Brien, L.G. and Wrigley, N. (1984) 'A generalised linear models approach to categorical data analysis: theory and applications in Geography and Regional Science', in G. Bahrenberg, M.M. Fischer and P. Nijkamp (eds) *Recent Developments in Spatial Data Analysis: Methodology, Measurement, Models*, Aldershot, Gower Publications, Chapter 14.

Odland, J. and Balzar, B. (1979) 'Localised externalities, contagious processes and the deterioration of urban housing: an empirical analysis', *Socio-Economic Planning Sciences* 13, 87–93.

O'Farrell, P.N. and Crouchley, R. (1984) 'An industrial and spatial analysis of new firm formation in Ireland', *Regional Studies* 18, 221–36.

Openshaw, S. (1983) '*The modifiable areal unit problem*: concepts and techniques in modern Geography', *Geo Abstracts* 10, Norwich.

Openshaw, S. and Goddard, J. (1987) 'Some implications of the commodification of information and the emerging information economy for applied geographical research in the UK', *Environment and Planning, A*

Openshaw, S. and Taylor, P.J. (1979) 'A million or so correlation coefficients: three experiments on the modifiable areal unit problem', in N. Wrigley (ed.) *Statistical Applications in the Spatial Sciences*, London, Pion.

Ord, J.K. (1972) *Families of Frequency Distributions*, London, Griffin.

Pahl, R. (1984) *Divisions of Labour*, Oxford, Blackwell.

Patten, J. (1976) 'Patterns of migration and movement of labour to three pre-industrial East Anglian towns', *Journal of Historical Geography* 2, 11.

Payne, C. (1977) 'The log-linear model for contingency tables', in C.A. O'Muircheartaigh and C. Payne (eds) *The Analysis of Survey Data*, London, Wiley.

Payne, C. (1979) 'Estimation of the parameters in the log-linear model using the usual constraints parameterisation', *GLIM Newsletter*, 27–9.

Payne, C.D. (ed.) (1986) *The GLIM System: Release 3.77*, Numerical Algorithms Group Ltd, Oxford, UK. Available from NAG Ltd, Mayfield House, 256 Banbury Rd, Oxford, OX2 7DE, or 1101 31st Street, Suite 100, Downers Grove, Illinois 60515–1263, USA.

Payne, S.L. (1951) *The Art of Asking Questions*, Princeton, Princeton University Press.

Piaget, J. (1952) *The Child's Conception of Number*, London, Routledge & Kegan Paul.

Pickles, A. (1986) *An Introduction to Likelihood:* Concepts and Techniques in Modern Geography, 42, Norwich, Geobooks.

Pindyck, R.S. and Rubinfeld, D.L. (1976) *Econometric Models and Economic Forecasts*, Tokyo, McGraw-Hill Kogakusha.

Pitman, E. (1936) 'Sufficient statistics and intrinsic accuracy', *Proceedings of the Cambridge Philosophical Society* 32, 567–79.

Plackett, R.L. (1974) *The Analysis of Categorical Data*, London, Griffin.

Pregibon, D. (1980) 'Goodness-of-link tests for generalised linear models', *Applied Statistics* 29, 15–24.

Pregibon, D. (1982) 'Score tests', in R. Gilchrist (ed.) *GLIM82: Proceedings of the International Conference on Generalised Linear Models*, New York, Springer-Verlag.

Presser, S. and Schuman, H. (1980) 'The measurement of a middle position in attitude surveys', *Public Opinion Quarterly* 44, 70–85.

Reese, P.A. and Richardson, M.G. (1984) 'GLIM3A', *GLIM Newsletter* 8, 13–24.

Reynolds, H.T. (1977) *The Analysis of Cross-classifications*, Glencoe, Free Press.

Rhind, D. (1983) 'System development problems in two major data handling packages: SASPAC and ACCESS', in D. Peuquet and J. O'Callaghan (eds) *Proceedings of the US/Australia Workshop on the Design and Implementation of Computer-based Geographic Information Systems*, International Geographical Union, Commission on Geographic Data Sensing and Processing.

Rhind, D. and Hudson, R. (1980) *Land Use*, London, Methuen.

Richards, M.G. and Ben-Akiva, M.E. (1975) *A Disaggregate Travel Demand Model*, Farnborough, Saxon House.

Ripley, B. (1981) *Spatial Statistics*, New York, Wiley.

Rosen, S. (1974) 'Hedonic prices and implicit markets: product differentiation in pure competition', *Journal of Political Economy* 82, 34–55.

Roy, S.N. and Kastenbaum, M.A. (1956) 'On the hypothesis of no interaction in a multi-way contingency table', *Annals of Mathematical Statistics* 27, 749–57.

Sayer, A. (1984a) 'Defining the urban', *Geojournal* 9, 279–85.

Sayer, A. (1984b) *Method in Social Science: A Realist Approach*, London, Hutchinson University Library.

Scheffé, H. (1959) *The Analysis of Variance*, New York, Wiley.

Searle, S.R. (1971) *Linear Models*, New York, Wiley.

Seber, G. (1977) *Linear Regression Analysis*, New York, Wiley.

BIBLIOGRAPHY

Senior, M.L. (1979) 'From gravity modelling to entropy maximising: a pedagogic guide', *Progress in Human Geography* 3, 175–210

Shenton, L.R. and Bowman, K.O. (1977) *Maximum Likelihood Estimation in Small Samples*, London, Griffin.

Sichel, H.S. (1982) 'Repeat buying and the generalised inverse Gaussian–Poisson distribution', *Applied Statistics* 31, 193–204.

Siegel, S. (1956) *Nonparametric Statistics*, New York, McGraw-Hill.

Silvey, S. (1975) *Statistical Inference*, London, Chapman and Hall.

Simon, G. (1974) 'Alternative analyses for the singly-ordered contingency table', *Journal of the American Statistical Association* 69, 971–6.

Skemp, R. (1971) *The Psychology of Learning Mathematics*, Harmondsworth, Penguin.

Smith, D.M. (1975) *Patterns in Human Geography*, Harmondsworth, Penguin.

Snee, R.D. (1982) 'Non-additivity and a two-way classification: Is it interaction or non-homogeneous variance?', *Journal of the American Statistical Association* 77, 515–19.

Sobel, K. (1980) 'Travel demand forecasting with the nested multinomial logit model', *Transportation Research Record* 775, 48–55.

Sprott, D.A. (1973) 'Practical uses of the likelihood function', *Selecta Statistica Canadiana*, vol. 1, Toronto, University Press Canada.

Stamp, L.D. (1948) *The Land of Britain, its Use and Misuse*, London, Longman.

Stapleton-Concord, C. (1984) 'A mover/stayer approach to residential mobility', *Tijdschrift voor Economische en Sociale Geografie* 75, 249–62.

Stetzer, F. (1976) 'Parameter estimation for the constrained gravity model: a comparison of six methods', *Environment and Planning, A* 8, 673–83.

Stevens, S.S. (1946) 'On the theory of scales of measurement', *Science* 103, 677–80.

Stevens, S.S. (1951) 'Mathematical models and psychophysics', in S.S. Stevens (ed.) *Handbook of Experimental Psychology*, New York, Wiley.

Stevens, S.S. (1959) 'Measurement, psychophysics and utility', in C.E. Churchman and R. Ratoosh (eds) *Measurement, Definitions and Theories*, New York, Wiley.

Swan, A.V. (1986) 'Introductory Guide', Part 1 in C.D. Payne (ed.) *The GLIM System: Release 3.77*, Oxford, Numerical Algorithms Group Ltd.

Taylor, P. (1977) *Quantitative Methods in Geography*, Boston, Houghton Mifflin.

Taylor, P. (1980) 'A pedagogic application of multiple regression analysis', *Geography* 203–12.

Thatcher, A.R. (1984) 'The 1981 Census of Population in England and Wales', *Population Trends* 36, 5–9.

Thrall, R.M., Coombs, C.H. and Davis, L.D. (1954) *Decision Processes*, New York, Wiley.

Timmermans, H. (1980) 'Consumer spatial choice strategies: a comparative study of some alternative behavioural spatial shopping models', *Geoforum* 11, 123–31.

Townsend, A.R. (1986) 'The location of employment growth after 1978: The surprising significance of dispersed centres', *Environment and Planning, A* 18, 529–45.

Townsend, A.R., Blakemore, M.J. and Nelson, R. (1987) 'The NOMIS database: availability and uses for geographers', *Area* 19, 43–50.

Train, K. (1986) *Qualitative Choice Analysis: Theory, Econometrics and Applications to Automobile Demand*, Cambridge, Mass., MIT Press.

Tukey, J.W. (1977) *Exploratory Data Analysis*, Reading, Mass., Addison-Wesley.

Uncles, M. (ed.) (1988) *Longitudinal Data Analysis: Methods and Applications*, London, Pion.

347

Unwin, D.J. (1981) *Introductory Spatial Analysis*, London, Methuen.

Upton, G. (1978) *The Analysis of Cross-tabulated Data*, Chichester, Wiley.

Upton, G. and Fingleton, B. (1979) 'Log-linear models in geography', *Transactions of the Institute of British Geographers* 4, 103–15.

Upton, G. and Fingleton, B. (1985) *Spatial Data Analysis by Example*, vol. 1, Chichester, Wiley.

Upton, G. and Fingleton, B. (1989) *Spatial Data Analysis by Example*, vol. 2, Chichester, Wiley.

Wald, A. (1943) 'Tests of statistical hypotheses concerning several parameters when the number of observations is large', *Transactions of the American Mathematical Society* 54, 426–82.

Watts, D. (1981) 'A task-analysis approach to designing a regression analysis course', *American Statistician* 35, 77–84.

Weber, N. (1981) 'A GLIM macro to test for marginal homogeneity in contingency tables', *GLIM Newsletter* 4, 28–30.

Wedderburn, R.W.M. (1974a) 'Generalised linear models specified in terms of constraints', *Journal of the Royal Statistical Society, B* 36, 449–54.

Wedderburn, R.W.M. (1974b) 'Quasi-likelihood functions, generalised linear models and the Gauss–Newton method', *Biometrika* 61, 439–47.

Wedderburn, R.W.M. (1976) 'On the existence and uniqueness of the maximum likelihood for certain generalised linear models', *Biometrika* 63, 27–32.

Wetherill, G.B. and Curram, J.B. (1984) 'The design and evaluation of statistical software for microcomputers', Applied Statistics Research Unit, University of Kent, Canterbury.

Wetherill, G.B. and Curram, J.B. (1985) 'The design and evaluation of statistical software for microcomputers', *The Statistician* 34, 391–427.

White, H. (1982) 'Maximum likelihood estimation of misspecified models', *Econometrica* 50, 1–25.

Whitehead, J. (1980) 'Fitting Cox's regression model to survival data using GLIM', *Applied Statistics* 29, 268–75.

Whittemore, A.S. (1978) 'Collapsibility of multidimensional contingency tables', *Journal of the Royal Statistical Society, B* 40, 328–40.

Wilkinson, G. and Rogers, C. (1973) 'Symbolic description of factorial model for analysis of variance', *Applied Statistics* 22, 392–9.

Williams, H. and Ortuzar, J. (1982) 'Behavioural theories of dispersion and the misspecification of travel demand models', *Transportation Research, B* 16, 167–219.

Williams, R. (1973) *The Country and the City*, London, Chatto & Windus.

Wilson, A.G. (1974) *Urban and Regional Models in Geography and Planning*, London, Wiley.

Withers, C.W.J. (1981) 'The geographical extent of Gaelic in Scotland 1698–1806', *Scottish Geographical Magazine* 97, 130–9.

Withers, C.W.J. (1984) *Gaelic in Scotland 1698–1981*, Edinburgh, Jon Donald.

Wrigley, N. (1976) *An Introduction to the Use of the Logit Model in Geography:* Concepts and Techniques in Modern Geography 10, Norwich, GeoAbstracts.

Wrigley, N. (1979) 'Developments in the statistical analysis of categorical data', *Progress in Human Geography* 3, 315–55.

Wrigley, N. (1980) 'An approach to the modelling of shop-choice patterns: an exploratory analysis of purchasing patterns in a British City', in D.T. Herbert and R.J. Johnston (eds) *Geography and the Urban Environment*, vol. 3, Chichester, Wiley.

Wrigley, N. (1981) 'Categorical data analysis', in N. Wrigley and R.J. Bennett (eds) *Quantitative Geography: A British View*, London, Routledge & Kegan Paul.

Wrigley, N. (1985) *Categorical Data Analysis for Geographers and Environmental Scientists*, London, Longman.

Wrigley, N. and Bennett, R.J. (eds) (1981) *Quantitative Geography: A British View*, London, Routledge & Kegan Paul.

Wrigley, N. and Dunn, R.J. (1984a) 'Stochastic panel data models of urban shopping behaviour: 1 Purchasing at individual stores in a single city', *Environment and Planning A* 16, 629–50.

Wrigley, N. and Dunn, R.J. (1984b) 'Stochastic panel data models of urban shopping behaviour: 2 Multistore purchasing patterns and the Dirichlet model', *Environment and Planning A* 16, 759–78.

Wrigley, N. and Dunn, R.J. (1984c) 'Stochastic panel data models of urban shopping behaviour: 3 The interaction of store choice and brand choice', *Environment and Planning A* 16, 1221–36.

Wrigley, N. and Dunn, R.J. (1985) 'Stochastic panel data models of urban shopping behaviour: 4 Incorporating independent variables into the NBD and Dirichlet models', *Environment and Planning A* 17, 319–31.

Wrigley, N., Guy, C.M., Dunn, R.J. and O'Brien, L.G. (1985) 'The Cardiff Consumer Panel: methodological aspects of the conduct of a long-term panel survey', *Transactions of the Institute of British Geographers* 10, 63–76.

349

INDEX

Adams, E. 35
Adjusted residuals 265
Agar, M. 19
Agnew, J. 300
Agresti, A. 286
Aitkin, M. 171, 175, 258, 259, 263, 295
Alden, J. 19
Aliasing 325
Altman, J. 19
Analysis of covariance 5
Analysis of deviance table 245, 261, 262
Analysis of variance 5, 175, 187,
   220–30, 246; assumptions 236–7;
   higher-order 227–30; one-way 220–7;
   random effects 230
Andersen, E.B. 5, 323
Anderson, E.W. 11, 17, 71, 124, 133,
   232
Arrow, K. 310
Aspatial data 20
Assymmetric log-linear models 277–8;
   and marginal totals 277
Atkinson, A. 182
Autocorrelation 91, 199, 200, 201, 203
Avadhani, M. 129

Baird, J.C. 35
Baker, R. 179, 182, 329
Balzar, B. 285
Barcharts 64
Barndorff-Nielsen, O. 161, 323
Bartholomew, D. 283
Bartlett, M. 253
Basic spatial unit 27
Bates, J. 305
Bateson, N. 11
Batty, M. 293, 294
Baxter, M. 294, 304, 325

Bayesian analysis 121, 271
Belsley, D. 182
Ben-Akiva, M. 310
Benedetti, J. 263
Bennett, R.J. 1, 296
Besag, J. 92
Bhapkar, V. 148
Bias 19, 126
Biased population 126
Bibby, J. 171
Birch, M. 248, 232, 253
Bishop, Y. 184, 232, 248, 249, 253, 271,
   272, 276, 308, 315
Blalock, H.M. 43, 82, 127, 154, 227,
   228, 230
Blaxter, M. 18
Blocking 305, 308
BLOGIT 311
Blundell, R. 311
BMDP 5, 249, 259, 318, 327
Boden, P. 12
Bowlby, S. 175
Bowman, K. 144
Boxplots 59–62, 75
Breen, R. 286, 287
Brodsky, H. 254–5
Broom, D. 285
Brown, M. 263
Brown, P. 215
Burn, C. 71
Burrough, P. 20
Buse, A. 184

Campbell, N.R. 35
Capture–recapture experiments 308–9
Cardiff consumer panel survey 11, 16,
   38, 46, 85, 113, 125, 296
Carruthers, A.W. 28

350

Printed in the United States
by Baker & Taylor Publisher Services